[江苏江宁汤山方山国家地质公园丛书]

走进汤山旅游指南

主　编	副主编	编写人员		
陶奎元	项长兴	陶奎元	项长兴	沈加林
	许汉奎	赵慕明	许汉奎	李皓亮

江苏江宁汤山方山国家地质公园

东南大学出版社
SOUTHEAST UNIVERSITY PRESS

图书在版编目（CIP）数据

走进汤山旅游指南/陶奎元主编. --南京：东南大学出版社，2013.3
（江苏江宁汤山方山国家地质公园丛书）
ISBN 978-7-5641-4101-1

Ⅰ. ①走… Ⅱ. ①陶… Ⅲ. ①地质-国家公园-旅游指南-南京市 Ⅳ. ①S759.93

中国版本图书馆CIP数据核字（2013）第027630号

编排设计：	王　辉　曹帅彪　何晓晴
摄　　影：	沈旻　王辉　李皓亮
	陶奎元　项长兴　叶正华
英文翻译：	马　彦　周　菊　戚建中

出版发行：	东南大学出版社
社　　址：	南京四牌楼2号　邮编：210096
出 版 人：	江建中
网　　址：	http://www.seupress.com
电子邮箱：	press@seupress.com
经　　销：	全国各地新华书店
印　　刷：	南通印刷总厂有限公司
开　　本：	889mm ×1194mm 1/24
印　　张：	5.5
字　　数：	325千字
版　　次：	2013年3月第1版
印　　次：	2013年3月第1次印刷
书　　号：	ISBN 978-7-5641-4101-1
定　　价：	28.00元

本社图书若有印装质量问题，请直接与营销部联系。
电话（传真）：025-83791830

丛书编委会

顾　问
王德滋

编委会主任
周育刚
刘震平
缪秀梅

编委会委员
周荣根
贝业海
张孝友
张　圣
韩　江
李　娟
叶正华

序 一

地质遗迹属于一种自然遗产，包括各种标准化石、典型的地层剖面、特征的地质构造以及具有保存价值的各种地质体。对于典型的地质遗迹应当倍加维护，因为它一旦被毁，将无法恢复。通过建设国家地质公园，能够将具有保存价值的地质遗迹有效地加以保护。

汤山方山国家地质公园位于南京市江宁区，它融自然景观与人文景点为一体，形成既具科学价值又具历史文化背景的旅游胜地。汤山方山国家地质公园由汤山和方山两个景区组成。汤山景区拥有典型的古生代—早中生代地层剖面，内含多种标准化石，系我国南方地层的立典之地。通过对典型地层剖面和化石的研究，可以勾画出南京地区的沧海桑田演变史。汤山葫芦洞内发现世界级的地质遗迹——南京猿人头盖骨化石，经研究属于直立人范畴。汤山还拥有驰名中外的汤山温泉和著名的明文化景点——阳山碑材。方山景区堪称天然的火山博物馆。站在中华门城堡上，向南遥望，平顶的方山清晰可见。方山是一座古火山锥，形成于距今1000万年以前，它是由炽热的（温度超过1000℃）岩浆喷至地表并快速冷凝为玄武岩堆积而成的盾形火山锥，可辨认出火山口的具体位置。方山景区还拥有著名的人文景点——定林寺和斜塔。

汤山方山国家地质公园是一个多功能的园区：一是高等院校本科生的教学实习基地；二是青少年的科普教育基地；三是融科学与人文为一体的地学旅游景区。汤山方山国家地质公园丛书的出版必将激发人们对学习地球科学知识的浓厚兴趣，并将增强人们保护自然环境的自觉性和积极性。

中国科学院院士
原南京大学副校长　王德滋

2012年10月

Foreword One

Geological heritage belongs to a kind of natural heritages, including a variety of standard fossils, typical stratigraphic sections, geological structures with the characteristics and various geological bodies worthy of preservation. Typical geological features should be doubly maintained, because once destroyed can they not be recovered. Through the construction of national geological parks, we will save the valuable geological heritages.

Located in Jiangning District of Nanjing, Tangshan-Fangshan National Geopark mixes natural landscape with cultural attractions. It is a tourist destination having both scientific value and historical cultural background. Tangshan-Fangshan National Geopark consists of two scenic districts—Tangshan and Fangshan. Tangshan Scenic District has a typical Paleozoic-Early Mesozoic stratigraphic section, containing a variety of standard fossils. It is the classical strata of southern China. The study of typical stratigraphic sections and fossils lays out the long history of the evolution of the Nanjing area. Nanjing ape-man skull fossils—world-class geological relics discovered in Hulu Cave in Tangshan—belong to Homo erectus by research. Tangshan also has Tangshan Hot Springs and famous Ming cultural attractions—Yangshan Tablet Material. Fangshan Scenic District is called the natural volcanic museum. Standing on the Zhonghuamen Castle and overlooking toward the south, you can see the flat-topped Fangshan. Fangshan, an ancient volcanic cone formed more than 10,000,000 years ago, was a shield volcanic cone made up of piled basalt formed when hot (temperatures in excess of 1,000℃) magma sprayed to the earth surface and then cooled rapidly. The specific location of the crater is identifiable. Fangshan Scenic District also has famous cultural attractions—Dinglin Temple and Dinglin Inclined Pagoda.

Tangshan-Fangshan National Geopark is a multi-purpose park: the undergraduate teaching internship base; youth science education base; Earth Science tourist attractions. The publication of the series of Tangshan-Fangshan National Geopark books is bound to stimulate interests in learning the scientific knowledge of the Earth, and enhance the consciousness and enthusiasm of the people to protect the natural environment.

Academician of Chinese Academy of Sciences
Former Vice President of Nanjing University

Wang Dezi

2012.10

序 二

　　地质公园是以山水景观为游览对象的一种新型公园，中国是世界上最先提出（1985年）、最早建成地质公园（2000年）的国家。地质公园具有保护地质遗产、传播地球科学知识、通过开展旅游促进地方经济社会可持续发展的重要功能。江苏是中国地质工作的发源地之一，地质条件多样，地质研究程度高，有建立各种类型地质公园的条件。现已建成江苏苏州太湖西山国家地质公园、江苏六合国家地质公园和江苏江宁方山汤山国家地质公园等三处，将来还会有更多的地质公园建立。游人走进地质公园不但可以欣赏山水美景，还可顺便获得山水由来的地球科学知识，以增添游兴。但是，山水形成的道理较为深奥，因此编写一部图文并茂、文字深入浅出的地质公园导游指南丛书，就显得非常迫切和必要了，江苏国家地质公园旅游指南就是为此而编撰的。

　　该丛书主要编写者陶奎元教授是资深地质学家，而且对地质公园建设有深入研究。他编写的这套丛书是我目前看到的最佳地质公园导游指南书。该书不但用图片、图解方式帮助游人直观地获得该公园山水景物形成演变的科学道理，还收入了许多"沧海桑田变迁"、地球物质组成等地学基础知识，以及该公园的地质研究史、文化史；为了帮助游人安排吃、住、行、游、购、娱活动，特别收入了详细的旅游咨询信息。该书不但能帮助到自助旅游的客人，还是中小学生，甚至大专院校师生学习地球科学知识的好读物。由于该书附有中英对照文字，还是向国外游客推介中国地质公园的好材料。

　　我怀着敬佩的心情向读者推荐这部难得的好书，希望读者喜欢它、使用它，并在使用过程中提出修改意见和建议，以便在再版时加以改进。我希望全国其他地质公园吸收该旅游指南的优点，结合实际编写出本地地质公园的旅游指南，以期更多的人游览地质公园，提高对地球科学的兴趣，用科学发展观作指导去游遍祖国的名山大川。

中国地质科学院研究员
中国旅游地学创始人
国家地质公园研究建设专家

2012.10.20于北京

Foreword Two

The geopark is a new kind of park in which the landscapes are the main tourism resources. China is the first country to propose (1985) and earliest constructed the geoparks in the world. The geoparks have many functions: the protection of geological heritages, the dissemination of knowledge of Earth Sciences and the important task to promote the sustainable development of local economy and society through developing tourism. Jiangsu Province is one of the cradles of the Chinese geological work with diverse geological conditions, a high degree of geological research, and the qualification of establishing various types of geological parks. Three national geoparks have been constructed in Jiangsu which are Xishan National Geopark in Taihu Lake of Suzhou, Luhe National Geopark and Tangshan-Fangshan Geopark. In the future, more geoparks will be constructed. Entering a geological park, visitors can not only enjoy the splendid views, but also gain the knowledge of Earth Sciences on the origin of landscapes which enrich the content of tourism. However, the knowledge on the origin of landscapes is too esoteric to understand which makes it very necessary and urgent to publish a series of illustrated geological park guidebooks written in simple words. Hence, the Travel Guide of Jiangsu Geoparks comes into being.

Prof. Tao Kuiyuan, the chief writer of this series, is a senior geologist and has in-depth study of the geopark construction.

This series of books is the best I've ever read, which not only uses pictures and graphic modes to help the visitors intuitively get the scientific basis of landscapes' formation and evolution, but also includes "earth-shaking changes", the basic knowledge of Geosciences such as earth materials composition, as well as the park's geological history, cultural history. What's more, in order to help visitors arrange food, lodging, visiting, shopping and entertainment activities, this series also includes detailed tourist advisory information. This book can not only help the independent travelers but also be good reading material for primary and secondary school students, and even the teachers and students of colleges and universities to learn the knowledge of Earth Sciences. Duo to its bilingual texts, it can also be good material for foreign tourists to promote the Chinese geoparks.

I recommend this rare good book to readers with admiration and I hope readers like it, use it, and make suggestions for revision which can be improved in the second edition.

I hope other geological parks across the country can absorb the advantages of this travel guide and write their own tourism guide according to the reality. More and more tourists are expected to visit Geoparks, raise interest in the Earth Sciences, and travel throughout the motherland with the guidance of the Scientific Outlook of development.

Researcher of China's Geological Science
Founder of China's Tourism Geology
Construction Expert of National Geological Park

Chen Anze

October 20th, 2012,
Beijing

前言

江苏汤山方山国家地质公园是拥有世界级、国家级的地质遗迹,景观丰富多样,历史文化淀积深厚,是处在省会城市中的一个地质公园。

建设地质公园的宗旨:其一是保护地质遗迹、保护生态环境,实施保护基础上开发、开发中保护的原则;其二是主动开展科学普及教育、环境友好教育,使公园成为社会大众喜爱科普的大教室;其三,发展旅游并带动地方社会经济发展。汤山方山国家地质公园丛书的出版在于推动科学建设地质公园,引导游客走进地质公园,实现寓学于游、寓教于游。

《走进汤山》旅游指南将带领你攀登、欣赏美丽的奇观,探索神秘的火山,追寻、品鉴悠久的文化历史。

Preface

Jiangsu Tangshan-Fangshan National Geopark in the provincial capital is a geopark which embraces a series of state-rank, even world-rank geologic remains, rich and variable landscapes, and deep accretion of culture over long history.

The goal to construct the geopark is as follows: 1. protect geologic heritage and ecologic environment, following the principle 'develop based on protection of the geopark, and protect in development of the park'; 2. actively carry out popular science education and environment friendly education to build the park into an open classroom for general public; 3. develop tourism to promote the local economy. Recent collection on Tangshan-Fangshan National Geopark to be published is expected to play a role in construction of the Geopark in a scientific way, to lead more tourists to get into the Geopark, and to realize 'learning from travel and educating within travel'.

The leaflet 'Enter the Fangshan' will accompany you to climb up and view the beautiful scenes, to search volcano for mystery, and to trace and savor the culture over long history.

汤山方山国家地质公园简介

Introduction to Tangshan-Fangshan National Geopark

汤山方山国家地质公园位于南京主城之东江宁区境内，面积29.15km²，由汤山与方山两个园区组成。通达便捷，属于处在城市中的地质公园。公园拥有丰富的地质遗迹景观和深厚的人文历史。公园主题是：

- 南京人祖宗、人类祖先—— 南京猿人（洞）；
- 古今闻名温汤圣水——汤山温泉；
- 1 000万年前神奇火山——方山；
- 守望600年皇家碑材——阳山碑材；
- 保存5亿多年来大地变迁、生物演化遗迹——地质走廊。

公园处于宁镇山脉之西段，留下几代地质学家的足迹与贡献，被誉为中国地质工作者的摇篮。

公园具有观光揽胜、休闲度假、科普旅游、地质实践、史迹追踪、娱乐美食等多种功能。

Tangshan-Fangshan National Geopark is situated in Jiangning District, east of Nanjing downtown with an area of 29.15km², which consists of Tangshan Scenic District and Fangshan Scenic District. As a city park, two scenic districts are well accessible. The geopark possesses rich geologic remains, landscapes and deep cultural heritage. The main topics of the park are as follows:

- The forefathers of Nanjing locals and ancestry of humans' —— Nanjing Ape-man;
- The holy water of Wentang Spring, famous for the ancient and present- Tangshan Hot Spring;
- The mystical volcano erupted 1,000 years ago —— Fangshan;
- The imperial tablet material retained for 600 years—— Yangshan Tablet Material;
- The earth change and organism evolution relics preserved for 500 million years—— Geologic Corridor.

The geopark is located at the western end of Ningzhen Ridge, which boasts of the reputation of 'the Cradle for China Geology'. There are footprints and contributions of prominent geologists of several generations.

The park has muliti functions as sightseeing, recreation and vacationing, science tourism, geological practice, history tracing, entertainment and cuisine.

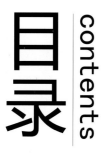

01 纵览汤山
Scanning of Tangshan

- 02 地理位置 Geographical Position
- 03 气候、水文 Climate and Hydrology
- 03 地形地貌 Terrain and Landform
- 05 地质概况 Geological Situation
- 08 地质研究历史 Geological Research History

11 南京猿人洞
Nanjing Ape-man Cave

- 12 南京猿人，人类祖先 Nanjing Ape-man, Human Ancestors
- 19 南京猿人之谜 Mystery of Nanjing Ape-man
- 20 葫芦洞景观奇特 Unique Landscape of Hulu Cave

25 阳山碑材
Yangshan Tablet Material

- 26 明文化村，古风遗韵
  Ming Villages, Antique Life
- 27 旷世碑材，史迹留痕
  Outstanding Massif, Historic Evidence
- 34 石灰岩层，科普园地
  Limestone Formation, Popular Science Field

39 汤山温泉
Tangshan Hot-spring

- 40 汤山温泉概况
  The General Situation of Tangshan Hot-spring
- 43 汤山温泉文化
  The Culture of Tangshan Hot-spring
- 46 名人与温泉有关的活动片段
  The Notables' Activities Related with Tangshan Hot-spring

47 旅游资讯
Tourist Information

- 48 主要景区
  Main Scenic spots
- 49 周边旅游
  Surrounding Tourism Areas
- 49 酒店
  Hotels
- 51 娱乐
  Recreation
- 52 旅游路线
  Tour Routes
- 53 附录
  Appendix
  汤山方山国家地质公园汤山园区导游图
  Tourist Map of Tangshan District, Tangshan Fangshan National Geological Park
- 54 汤山园区地层简表
  Strata Profile of Tangshan District

汤山园区简介

汤山园区是汤山方山国家地质公园园区之一。面积21.05平方千米,包括汤山、阳山、孔山、湖山。

园区具有世界级意义、距今64万~56万年的南京猿人(洞);600年前造就的天下第一碑——阳山碑材;千年流淌的汤山温泉;5亿多年地质历史的地质剖面走廊。

汤山园区已建有南京猿人洞景区、阳山碑材景区以及水上乐园、温泉度假区以及博物馆区等。

汤山集洞之奇、泉之韵、碑之最于一体,具有观光探奇、科普体验、地质教学、休闲度假、沐浴养生、娱乐美食的旅游目的地。

Introduction to Tangshan Scenic District

Tangshan Park, one of the Geopark's scenic districts with an area of 21.05km², includes Tangshan, Yangshan, Kongshan and Hushan.

The park has the world-rank Nanjing Ape-man Cave with a history of 560,000-640,000 years. The park has the "No.1 Tablet in the world" built 600 years ago, Yangshan Tablet Material. The park has Tangshan Hot-spring flowing for thousands of years. The park also has 500 million-year Geologic Corridor.

Tangshan Park has built the scenic spots of Nanjing Ape-man Cave, Yangshan Tablet Material, Water Park, Hot-spring Resort and Museum area.

Tangshan sets the strange cave, the rhythmic spring and the No.1 tablet in a body. So it has multi functions as sightseeing, science tourism, geological practice, vacationing, health bathing, entertainment and good food.

纵览汤山
Scanning of Tangshan

01

地理位置

汤山方山国家地质公园汤山园区，位于江宁区北部，隶属汤山镇街道，距中山门30千米。汤山处在快速发展的城市之内。汤山是南京通往镇江、常州、无锡、苏州、上海的门户。沪宁高速、沪宁高铁、沪宁城际、宁杭高速等至汤山园区都有出入口。并建有南京三环于汤山园区东缘，南京城区至汤山轻轨、快速通道正在建设中。现今已有6条公交线路进出汤山。

Geographical Position

Tangshan Park is situated in the north of Jiangning District, belonging to the Tangshan town street. It is 30km away from Zhongshanmen, Nanjing. Tangshan is located in the rapidly developing city. It is the Nanjing's door to Zhenjiang, Changzhou, Wuxi, Suzhou and Shanghai. The Shanghai-Nanjing highway, the Shanghai-Nanjing high-speed rail, the Shanghai-Nanjing inter-city train, the Nanjing-Hangzhou highway and so on, all have entrances to Tangshan Park. Nanjing Tri-circle is to the east of the park, and the light rail from Nanjing city to Tangshan is under the construction. Until now, there have been 6 bus lines to Tangshan.

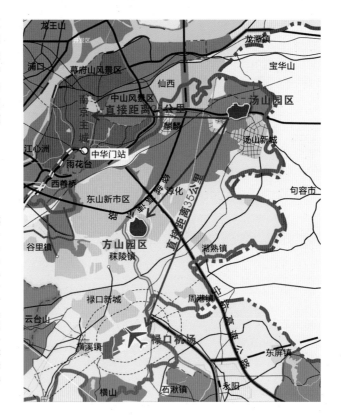

气候、水文

气候：汤山地处亚热带季风气候区，温暖湿润，四季分明，年平均气温15℃，无霜期237天。1月是全年最冷月，平均气温2.3℃。夏季受副热带高压影响，7月是全年最热月，平均气温27.9℃。年极低气温-13.3℃（1977年1月31日），年极高气温40.7℃（1959年8月22日）。

年平均降水量1 060毫米，主要集中在6～8月，约占全年降水量的50%以上。年蒸发量1 400～1 500毫米。

水文：园区内有七乡河、九乡河、汤水河三条季节性河流。七乡河、九乡河向北流入长江，汤水河向南流入秦淮河。汤泉湖容量275万立方米。

Climate and Hydrology

Climate: Tangshan belongs to the northern subtropical monsoon climate zone. It is characterized by mild temperature, distinctive seasons. The annual average temperature is 15℃. The frost-free period has 237 days. January is the coldest month in a year with the average temperature of 2.3 ℃. Influenced by the subtropical high pressure, July becomes the hottest month with the average temperature of 27.9 ℃. The annual lowest temperature was -13.3 ℃(1,31,1977),and the annual highest temperature was 40.7 ℃(8,22,1959).

The annual average precipitation is 1,060mm, mainly from June to August, which accounts for 50% of the total annual precipitation. The annual evaporation is 1,400 ～1,500mm.

Hydrology: Tangshan has three seasonal rivers—Qixiang River, Jiuxiang River,Tangshui River. Qixiang River and Jiuxiang River flow north into the Yangtze River while Tangshui River flows south into Qinhuai River. Tangquan Lake has a capacity of 2,750,000m³.

地形地貌

汤山地区属低山丘陵区，汤山山体走向近东西，中部高、两头低。主峰汤山（团子尖）标高292.8米，西部汤山头标高241米，东部雷公山标高142米。山峰间鞍部较低，整个山体形似元宝。孔山标高342米。阳山标高239.7米。湖山标高172米。

Terrain and Landform

Tangshan belongs to the low mountains and hills. The Tangshan Mountain sets a west-east trend, which is high in the center and low in both ends. The main peak,Tangshan(Tuanzijian)'s elevation is 292.8m. Tangshan Head's elevation is 241m in the west. Leigongshan Mountain's elevation is 142m in the east. The whole saddle of the mountain is low, so the mountain looks like Chinese Yuanbao (shoe-shaped gold ingot). Kongshan's elevation is 342m. Yangshan's elevation is 239.7m. Hushan's elevation is 172m.

汤山方山国家地质公园（汤山）地质图

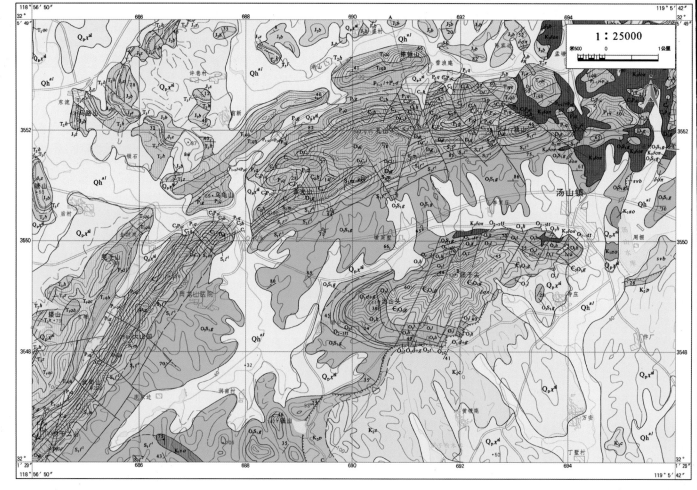

地质剖面图

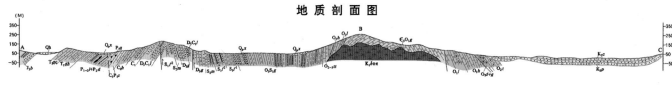

图 例

一. 地层

Qh	全新统
Q$_p$s^{xl}	下蜀组
K$_2$c	赤山组
K$_2$p	浦口组
svb	上党火山岩
J$_3$b	北象山组
J$_1$z	钟山组
T$_3$f	范家塘组
T$_2$h	黄马青组
T$_2$z	周冲村组
T$_1$qc	青龙组沧波门段
T$_1$qh	青龙组湖山段
P$_2$d	大隆组
P$_{1-2}$l	龙潭组
P$_1$g	孤峰组
P$_1$q	栖霞组
C$_2$P$_1$c	船山组
C$_2$h	黄龙组
C$_1$	下碳统（老虎洞组、和州组、高骊山组、金陵组）
D$_3$C$_1$l	擂鼓台组
D$_3$g	观山组
S$_2$m	茅山组
S$_2$f^2	坟头组上段
S$_2$f^1	坟头组下段
O$_3$S$_1$g	高家边组
O$_{2-3}$ll	汤头组
O$_2$l	汤山组
O$_1$t	牛潭组
O$_1$d	大湾组
O$_1$h	红花园组
O$_1$l	仑山组
∈$_3$O$_1$t	观音台组
fml	混杂岩（时代未定）

其他符号：
- tcs 钙质构造岩

二. 岩浆岩

1. 次火山岩
- K$_1$π 石英粗面岩
- K$_1$αo 石英安山岩
- K$_1$δoπ 石英闪长斑岩

2. 脉岩
- δμ 闪长玢岩
- δoπ 石英闪长斑岩

三. 其他

- 逆断层
- 逆掩断层
- 正断层
- 平移断层
- 实测及推测性质不明断层
- 实测及推测地质界线
- 不整合地质界线
- 正常地层产状及倾角
- 倒转地层产状及倾角
- 流面产状及倾角

地质概况

汤山、阳山（孔山）、湖山处于宁镇山脉西段，由5亿年的寒武系至距今1.76亿年的三叠系地层组成近东西向的背斜、向斜复式褶皱，伴随发育多组区域性断裂。

晚侏罗系、白垩系有小型的石英闪长斑岩岩体、岩脉侵入。

宁镇山脉地质构造以褶皱为主，断层为辅，形成宁镇山脉大致东西走向的褶皱构造，由北向南形成3个复背斜和2个复向斜，汤山即位于最南部的汤（山）—仑（山）复背斜的西段。

Geological Situation

Tangshan, Yangshan(Kongshan), Hushan are all located in the western part of Ningzhen Mountains. Their west-east anticlines are composited by the stratum of the 500 million years' Cambrian system to the 176 million years' Trias system, with syncline double fold, following with many sets of regional fractures.

In late Jurassic, Cretaceous there was the invasion of small quartz diorite porphyry rock mass and vein.

The geological structure of Ningzhen Mountains gives priority to folding, supplemented by fault, forms Ningzhen Mountains' fold structure from east to west, and forms three anticlinoriums and 2 synclinoriums from north to south. Tangshan is located in the western part of the southernmost point's Tangshan-Lunshan anticlinorium.

1. 地质年代

根据古生物化石、同位素方法测定岩石的年龄，确定岩石地层年代，并分为代、纪、世、期、时。如奥陶纪、志留纪等。

2. 褶皱构造——背斜、向斜

岩石通常呈层状，称为岩层。当地壳运动，受力的作用岩层扭曲成各种形状，岩层向上或向下弯曲。向上弯曲称为背斜，向下弯曲称为向斜。

1. Geologic Time

According to ancient fossils, isotope method, we can determine the age of the rock and determine the stratigraphic time of the rock, and divide them into era, period, epoch, stage age, age Ma, such as Ordovician period, Silurian period, etc.

2. Fold Structure——Anticline and syncline

Rock is usually stratified, known as rock stratum. When the earth crust moves, affected by the stress, rock stratums distort into various shapes, sweeping up or bending down. When the rock upsweeps, it is called anticline; when the rock bends down, it is called syncline.

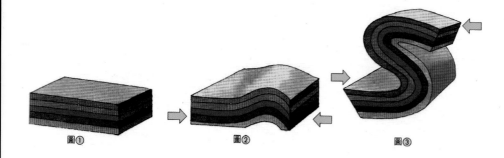

图① 图② 图③

3. 断层
3. Fault

断层类型 Fault type

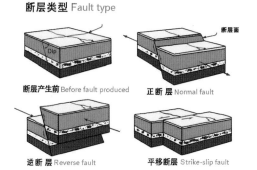

断层产生前 Before fault produced
正断层 Normal fault
逆断层 Reverse fault
平移断层 Strike-slip fault

宁镇山脉——汤山地区地质年代与地层

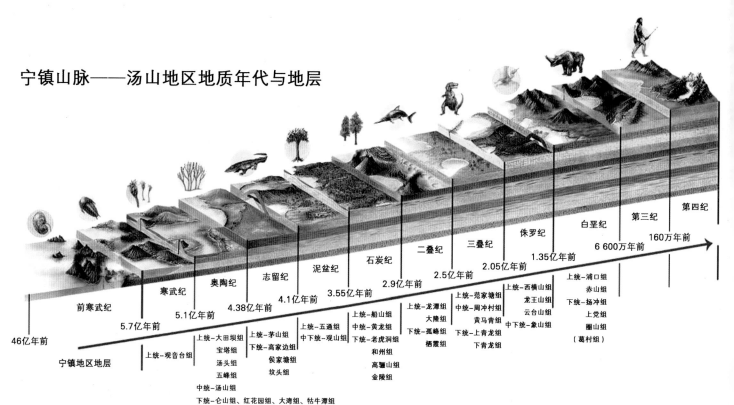

地质研究历史

宁镇山脉地层发育较齐全，岩浆岩类型多，地质构造现象颇为典型，是地质考察与教学实习的良好场所。宁镇山脉被称为地质工作者的摇篮，许多地质界前辈为研究宁镇山脉地质作出了杰出贡献。

早在1917年，叶良辅、丁文江先后分别调查宁镇山脉等地区地质构造。丁文江于1919年编著出版《扬子江下游之地质》一书，是中国人最早系统全面研究苏南地区地质构造及岩浆活动等的论著。

1920—1922年，刘季辰、赵汝钧对江苏全省地质矿产情况进行调查，于1924年编著出版《江苏地质志》，对全省地层、岩浆活动等均有详细论述，对三叠系及其以后的地层划分有建树，首次建立了侏罗系至白垩系的地层层序。许多地层名称沿用至今。

1920—1934年，时任中央地质调查所所长李四光组织，指导李捷、李毓尧、朱森等人分别对宁镇山脉东西段、茅山山脉的地层、构造进行地质调查（火成岩由叶良辅、喻德渊负责），调查成果一书《宁镇山脉地质》于1935年出版。书中建立了宁镇山脉古生界至新生界的地层层序，共建24个地层单位。大多数地层的时代和名称沿用至今。

1928年，谢家荣、张更对汤山地区的地层、温泉等进行调查，发表了《南京汤山及其附近地质》，绘制了汤山温泉之分布图。同年谢家荣的《南京钟山地质与首都之井水供给关系》发表。

1930年，李四光、朱森先后到栖霞、龙潭进行地质调查，绘制了五千分之一栖霞山地质图，发表了《栖霞灰岩及其相关地层》（1930）、《南京龙潭地质指南》。还有许杰、俞建章、翁文灏、孙云铸、杨钟健、穆恩之等诸多老一辈地质工作者对宁镇山脉地质进行专题调查研究，取得很多重要成果。

李四光
（1889--1971）

丁文江
（1887--1936）

翁文灏
（1889--1971）

李毓尧
（1894--1966）

Geological Research History

The formation development of Ningzhen Mountains is complete. There are many magmatic rock types, and geological structure is a typical phenomenon. It is a good place for geological investigation and practice teaching. Ningzhen Mountains are called the cradle of geological workers. Many geological predecessors have made outstanding contributions to the Ningzhen Mountains' research.

Early in 1917, Ye Liangfu, Ding Wenjiang have made investigation of Ningzhen Mountains' geological structure respectively. Ding Wenjiang's The Geology of Yangtze River Downstream, written and published in 1919, is the first Chinese book that systematically and comprehensively studied the geological structure and the magmatic activities of southern Jiangsu.

In 1920—1922, Liu Jichen and Zhao Rujun conducted a survey of the geology and mineral resources of Jiangsu Province. They wrote the book The Annals of Jiangsu Geology which was published in 1924. This book presents the province's stratums and magmatic activities in detail, making achievements on the stratigraphic division of the Triassic and later times, founding the stratigraphic sequence from the Jurassic to the Cretaceous for the first time. Many formation names are used up to now.

In 1920—1934, Li Siguang, the director of the Center Address Survey at that time, guided Li Jie, Li Yuyao, Zhu Sen etc. on the geology survey of Ningzhen Mountains' eastern and western parts, the stratums and formation of Maoshan Mountains (Ye Liangfu, Yu Deyuan were in charge of igneous rock). The survey results Ningzhen Mountains' Geology was published in 1935. In this book, the stratigraphic sequence of Ningzhen Mountains' Paleozoic erathem to Cenozoic group is founded, and there are totally 24 stratigraphic units. Most of stratigraphic ages and names are used up to now.

In 1928, Xie Jiarong, Zhang Geng conducted a survey of Tangshan area's stratums and hotsprings. They published The Geology of Nanjing Tangshan and Nearby. In the same year, The Relationships between Nanjing Zhongshan Geology and the Capital's Well Water Supply was published, which was written by Xie Jiarong.

In 1930, Li Siguang and Zhu Sen came to Qixia Mountain and Longtan. They mapped the geological map of Qixia Mountain by 1:5,000 and published Qixia Limestone and Related Formations(1930), Geological guide of Nanjing Longtan. Many elders of geological workers, such as Xu Jie, Yv Jianzhang, Weng Wenhao, Sun Yunzhu, Yang Zhongjian, Mu Enzhi, have done a lot of project researches in the geology of Ningzhen Mountains and have gotten a lot of important achievements.

李捷、李毓尧、朱森三人考察照片

南京大学地质系、中国地质调查局南京地质调查中心（南京地质矿产研究所、地质矿产部原华东地质矿产研究所）、中科院、南京地质古生物研究所、江苏省区测队（现江苏省地质调查院）、南京师范大学地理系等单位，对本地区的地层古生物、岩石、构造诸多方面进行了大量的专题研究，发表了很多论文和专著。

Many departments have done a lot of researches in the regional stratigraphic paleontology, rock and geological structure, such as the Geology Department of Nanjing University, the China Geological Survey's Geological Research Center of Nanjing (Nanjing Institute of Geology and Mineral Resources, the National Geology and Mineral Ministry's Former East China Institute of Geology and Mineral Resources),Chinese Academy of Sciences,Nanjing Institute of Geology Paleontology,Jiangsu Provincial Test Team (now Geology Survey Institute of Jiangsu Province),the Geography Department of Nanjing Normal University. They have published many related theses and academic monographs.

中国地质学的奠基人（1933年夏）中多位在宁镇山脉作过研究

前排左起：章鸿钊、丁文江、葛利普、翁文灏、德日进
中排左起：杨钟健、周赞衡、谢家荣、徐光熙、孙云铸、谭锡畴、王 文、尹赞勤、袁复礼
后排左起：何作霖、王恒升、王竹泉、王曰伦、朱焕文、计荣森、孙健初

汤山猿人洞景区是汤山方山国家地质公园主要景区之一。
主题：寻找人类祖先，拜访南京猿人，观赏溶洞奇观。

Tangshan Ape-man Cave is one of Tangshan-Fangshan National Geopark's main scenic spots.

Themes: Looking for human ancestors, visiting Nanjing Ape-man, appreciating cave wonders.

南京猿人洞
Nanjing Ape-man Cave | 11

南京猿人,人类祖先

Nanjing Ape-man, Human Ancestors

1、南京猿人一号头骨是位少妇

南京猿人一号头骨长16厘米,宽13厘米,脑量860毫升(男猿人在1 000毫升左右,现代人为1 400~1 700毫升),颅骨表面纤细,光滑,故应为女性。根据她的上颌骨第二前臼齿的齿槽看,该牙齿根高仅为13.5毫米,齿根近中远中径5.2毫米,这与北京猿人女性的齿根相应值分别为13.6~16.2毫米和5.3~5.8毫米很接近。根据她牙齿(臼齿)骨片缝合状况,推测为21~30岁。

1. The Nanjing Ape-man's No.1 Calvarium Belonging to a Young Woman

The Nanjing ape-man's No.1 calvarium is 16cm long, 13cm wide, with 860ml cranial capacity (while a male ape-man's cranial capacity is 1000ml, a modernist's is 1 400~1 700ml). The surface of the skull is fine and smooth, so it should be from a female. According to her second premolar's tooth space of the upper jaw bone, we can see the tooth root is only 13.5mm tall; the crown angulations are 5.2mm. They are close to the female ape-man's corresponding numeric of 13.6~16.2mm and 5.3~5.8mm. She is guessed to be around 21~30 years old according to the suture status of her teeth bone piece.

南京猿人一号头骨复原头像

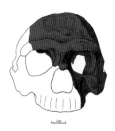

南京一号头骨复原,前面观

南京一号头骨复原,左侧观

南京一号头骨复原,顶面观

复原后的头骨前侧面观

南京一号头骨枕骨-左顶骨块
枕面观和脑面观

南京一号头骨颅窝-面骨块
顶面观

南京一号头骨颅窝-面骨块
后面观

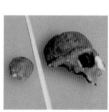

头骨前面观

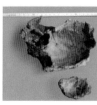

头骨内面观

2、南京猿人二号头骨是个壮年汉子

南京猿人二号头盖骨其颅骨粗壮，骨壁厚重，颅腔宽阔，表明他应为男性。

根据颅骨外矢状缝和冠状缝愈合程度推测他的年龄在24~41岁之间。若考虑到头骨小者愈合较早，推测他的年龄应在30~40岁之间。

南京猿人处在由南方古猿至智人的过渡阶段——直立人。

3、南京猿人一号头骨的时代

葫芦洞内南面的小洞（即猿人洞）里，堆积着1.4米厚的红褐色的粘土层。粘土层上面覆有2~3厘米厚的钙板。钙板上有石笋。钙板下面的粘土层中有许多哺乳动物化石及南京猿人一号头骨、孢粉化石和植硅体。科学家们通过对钙板中放射性元素铀的测量，得知其年代约距今56万年。动物化石是在钙板层之下，其年代应当比钙板层更早。而从动物化石氨基酸外消旋年代的测定，确定它们是距今60多万年的产物。从而推断出南京猿人一号头骨的距今年代为56万~64万年。

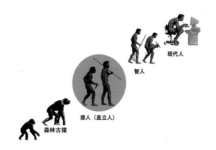

2. The Nanjing Ape-man's No.2 Calvarium Belonging to a Mature Man

The Nanjing ape-man's No.2 calvarium is stout with a thick bone wall and a broad cranial cavity, so it should be from a male.

According to the calvarium's sagittal suture and coronary joint degree of healing his age can be inferred between 24~41 years old. If considering the small skull healing earlier, we can assume that he should be in the age of 30~40.

Nanjing ape-man was in the transition stage of australopithecus to homo sapiens—homo erectus.

3. Time of the Nanjing Ape-man's No.1 Calvarium

There has an accumulation of 1.4m thick rufous clay layer in the small cave (Apeman Cave) to the south of Hulu Cave. The clay layer is covered by a flowstone layer of 2~3cm thick. There are stalagmites on the flowstone layer. A lot of mammal fossils, the Nanjing ape-man's No.1 calvarium, sporopollen fossils and phytolith are in the clay layer below the flowstone layer. Scientists measured the radioactive element Uranium in the flowstone layer and inferred its time—about 560,000 years ago. Animal fossils are under the flowstone layer, so their times were earlier than the flowstone layer's. And according to the measurement of the times of animal fossils' amino acid racemes, they are assured to be the products more than 600,000 years ago. So we can infer that the time of the Nanjing ape-man's No.1 calvarium was 560,000 ~ 640,000 years ago.

南京猿人及其遗址综合研究项目组的部分成员考察葫芦洞南侧小洞剖面现场

南京汤山早期人类文化遗址综合研究专家组部分成员考察葫芦洞

4、南京猿人在洞内发现的位置

猿人小洞在大洞内南面洞底下3~4米处，小洞面积6~7平方米，洞高2.3米，洞内沉积的红褐色黏土层1.1米厚。

在靠近下部的土层中含有许多哺乳动物化石。南京猿人一号头骨就产于此。南京猿人二号头骨产于小洞与大洞过道处地层中，年代比一号头骨晚20多万年。

4. The Position of Nanjing Ape-man Discovered in the Cave

The little ape-man cave is 3~4 meters below the south bottom of the big cave and its square is 6~7m²;its height is 2.3m. The rufous clay layer deposited in the cave is 1.1m thick.

There are many mammal fossils in the soil layer near the bottom. The Nanjing ape-man's No.1 calvarium was discovered here. The Nanjing ape-man's No.2 calvarium was discovered in the soil layer of the passageway between the little cave and the big cave and its time was more than 200,000 years later than the Nanjing ape-man's No.1 calvarium.

发现南京猿人一号头骨的土层

发现动物化石的土层

5、洞穴内发现大量的哺乳动物化石

在南京汤山葫芦洞内，共发现两个哺乳动物的化石层位。一层为大洞动物群，共5目13科16属17种。它们是：马铁菊头蝠、鼠耳蝠（未定种）、变异仓鼠、根田鼠、似小林姬鼠、棕熊、黑熊、似北方豺、南方猪獾、中国鬣狗、梅氏犀、李氏野猪、葛氏斑鹿、肿骨鹿、毛冠鹿、狍（未定种）、似德氏水牛。

5. A Lot of Mammal Fossils were Discovered in the Cave

Two mammal fossils horizons were discovered in the Hulu Cave of Nanjing Tangshan. One is the big cave's fauna, totally 5 orders, 13 families, 16 genuses and 17 species. They are: Rhinolophus ferrumequinum, Myotis sp., Cricetinus varians, Microtus oeconomus, Apodemus cf. syhaticus, Ursus arctos, Ursus thibetanus, Cuon cf. alpinus, Arctonyx collaris, Pachycrocuta sinensis, Dicerorhinus mercki, Sus lydekkeri, Cervus (Sika) grayi, Megaloceros pachyosteus, Elaphodus cephalophus, Capreolus sp., Bubalus cf. teilhardi.

第二层位产于猿人小洞内，与一号头骨伴存，共4目11科14属16种，称之为小洞动物群。它们是：棕熊、黑熊、中国鬣狗、虎、豹、中华貉、狐（未定种）、南方猪獾、李氏野猪、肿骨鹿、葛氏斑鹿、水牛（未定种）、梅氏犀、马（未定种）、剑齿象（未定种）、小型鹿属种未定。

此处驼子洞内也发现丰富的哺乳动物化石。

The other is the little cave's fauna in the little cave, together with the No.1 calvarium, totally 4 orders, 11 families, 14 genuses and 16 species. They are: Ursus arctos, Ursus thibetanus, Pachycrocuta sinensis, Panthera tigris, Panthera pardus, Nyctereutes sinenisis, Vulpes sp., Arctonyx collaris, Sus lydekkeri, Megaloceros pachyosteus, Cervus (Sika) grayi, Bubalus sp., Equus sp., Stegodon sp., Cervidae gen. et sp. Indet.

Many mammal fossils were also discovered in Tuozi Cave here.

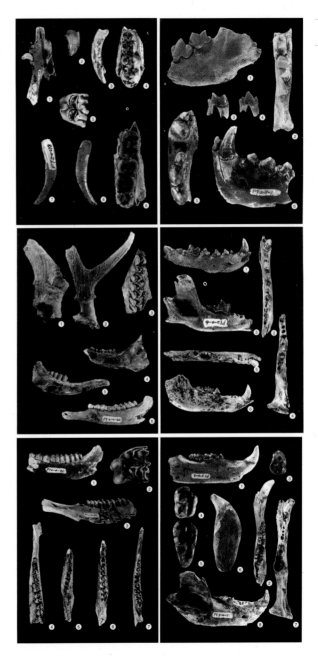

6、南京猿人与北京猿人比较

南京猿人与北京猿人有许多相似，如头骨尺寸均较小，头颅低而狭长，脑量在860至915毫升左右，前额均低平、后倾，颅盖低，最大宽度约近耳处，骨壁均在10毫米左右（现代人仅5毫米），年代距今都在60多万至20万年间，且晚期的猿人都比较进化等。但他们也有区别，如南京猿人枕骨轮廓线弯度小，不明显后突，也不是发髻状，头骨宽度相对长，鼻梁骨向前突而高耸，上颌额骨突有丘状膨隆结构，颜面上部扁平度较高，颜面纵向突度强，面部更低而相对较宽等，这说明我国北方人和南方人在猿人时代就有差别了。

6. Comparison of Nanjing Ape-man and Beijing Ape-man

There are many similarities between Nanjing Ape-man and Beijing Ape-man, such as smaller skull sizes; low, long and narrow heads; cranial capacities of about 860~915ml; flat and sloping foreheads; low skull caps, maximum widths near to ear places; all around 10mm thick bone walls (while modernists only 5mm thick); more than 200,000 to 600,000 years' ages; later ape-men evolved well and so on. But they have differences too, for example, Nanjing Ape-man's occipital profile bending is small, neither obviously backward protruding, nor bun shaped. The skull width of Nanjing Ape-man is relatively long. Their nasal bones are projecting forward and towering. Their maxillary frontals protrude dome-shaped bulging structures. The flat degree of their facial upper parts is higher, and their facial longitudinal protruding degree is also high. Their faces are lower and wider relatively. All of these show that northerners and southerners of our country have had differences since the ape-man's age.

南京猿人（左）与北京猿人（右）

7、南京猿人在人类进化中的位置
7. Nanjing Ape-man's Position in the Human Evolution

南京猿人在人类从猿到人的进化中处于在南方古猿到智人之间直立人的位置，可以直立行走。

In the human evolution history from ape to human, Nanjing Ape-man was in the transition stage of australopithecus to homo sapiens—homo erectus. They could walk upright.

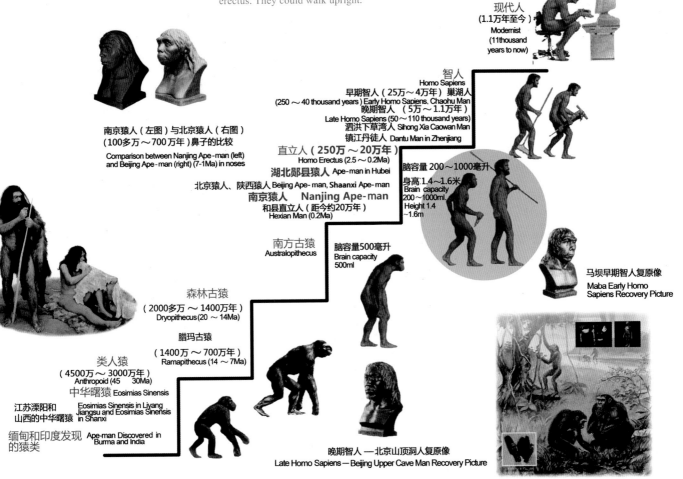

南京猿人在人类进化中的位置
Nanjing Ape-man's Position in the Human Evolution

8、南京猿人发现的意义

1、丰富了我国猿人的化石宝库。北京周口店先后发现了6个头盖骨,但5个已在抗日战争中遗失了,故南京猿人头骨化石更显得弥足珍贵。

2、改变了我国江南无猿人化石的历史。南京猿人是首次在长江以南发现,尤其她(他)与北京猿人恰好南、北遥相呼应,就更引人瞩目。这对研究我国古人类和动物群的迁徙及当时古气候、古地理和古环境都极为重要。

3、为人类发源地的探索提供了依据。人类起源地的两种论点至今争论很大,一是人类起源于非洲的"非洲论",一是人类起源"多地论"。南京猿人的发现,为多地论提供了更多依据。

4、把江苏地区古人类历史前移了50多万年。

5、提升了南京地区文化品位。世界各地发现的猿人化石地点,大部分位于偏远的山区,而南京猿人位于南京的近郊。

8. The Significance of Discovering Nanjing Ape-man

1. It enriches our country's treasure house of ape-man fossils. Though six skulls were discovered in Zhoukoudian, Beijing, five were lost during the Anti-Japanese War. So that is why Nanjing Ape-man's skull fossils are so precious.

2. It changs our history of no ape-man fossil in South China. It is the first time that ape-man was discovered in the south of the Yangtze River. She (He) and Beijing Ape-man echoed each other at a distance, which made more eye-catching. It is important to the study of China's ancient human and fauna migration and the ancient climate, ancient geography and ancient environment.

3. It provides a basis for the exploration of the human source. The two theories of the human source have been debated to this day. One is Africa's theory—humans originated in Africa; the other is Many Places' theory. The discovery of Nanjing Ape-man provides more evidence to Many Places' theory.

4. It puts forward the history of ancient human in Jiangsu area to more than 500,000 years ago.

5. It improves the cultural taste of Nanjing area. Most of the discovery places of ape-man fossils in the world are in remote mountain areas, while Nanjing Ape-man is located in the suburb of Nanjing.

出席南京汤山早期人类文化遗址综合研究专家组第一次会议的专家组成员及江苏省、南京市有关领导。前排左起:纪仲庆、佘之祥、蒋赞初、李星学、吴汝康、张怀西、徐京安、王湛、张连发、李根章、吴贻范、黄瑞骥、魏正瑾;后排左起:易家胜、张宏、钟石兰、徐钦琦、李士、梁白泉、穆西南、许汉奎、刘泽纯、穆道成、刘金陵、张银运。

由吴汝康、李星学等人编写的《南京直立人》

李星学院士在观察"猿人小洞"和地层

南京猿人发现经过

1990年3月22日在汤山北坡放炮采石时炸出一个小洞——次日采石工人下洞才发现洞很大且有大量动物化石。1993年3月13日清淤土时挖出一个头骨,经专家确认为猿人一号头骨。接着在大洞与小洞之间发现猿人二号头盖骨,1993年3月28日新华社报道了这一重大发现。

The Discovery Process of Nanjing Ape-man

In March 22,1990, a small cave was fried out when workers quarried in the northern slope of Tangshan. The next day they found out a lot of animal fossils in the cave. In March 13,1993, a skull was dug out when people were cleaning silt soil. The experts confirmed it was Nanjing Ape-man's No.1 calvarium. Then Nanjing Ape-man's No.2 calvarium was discovered between the big cave and the small cave. Xinhua News Agency reported this important discovery in March,28,1993.

南京猿人之谜
Mystery of Nanjing Ape-man

1、南京猿人牙齿之谜

北京大学吕遵锷教授称在葫芦洞的猿人小洞发掘出一枚猿人的臼齿;但中国科学院古脊椎动物与古人类研究所的专家们一致认为那是一枚智人的臼齿。若是智人臼齿,它怎么会出现在小洞的老地层中?

1.Mystery of Nanjing Ape-man's teeth

Lv Zun'e, a professor from Beijing University, said an ape-man's molar was dug out in the small cave of Hulu Cave. But the experts from the Institute of Vertebrate Paleontology and Paleoanthropology Chinese Academy of Sciences agreed that it was a homo sapien's molar. If it was a homo sapien's molar, how could it appear in the old strata of the small cave?

2、南京猿人生病之谜

南京猿人的1号头骨有严重的骨膜炎,这是世界上首次在猿人中发现疾病。但为何生病?是否与她死亡有关?现仍是个谜。

2.Mystery of Nanjing Ape-man's illness

The Nanjing Ape-man's No.1 calvarium shows serious periostitis, so it is the first time in the world that people have discovered ape-man's illness. But why she fell ill and whether the illness caused her death remain mysteries.

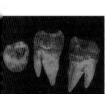
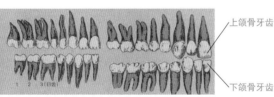

现代人牙齿　　　北京猿人牙齿

上颌骨牙齿
下颌骨牙齿

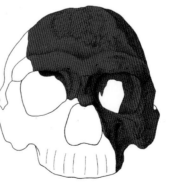

前面观　　　顶面观

3、南京猿人高鼻子之谜

南京猿人鼻骨高耸,是由于为适应寒冷气候所致,还是与欧洲人基因交流的结果,目前专家还在争论中。

3.Mystery of Nanjing Ape-man's High Nose

Nanjing Ape-man had a high nasal bone. Now experts are still arguing that whether it is the result of adapting to the cold climate, or gene exchanging with Europeans.

4、南京猿人2号头骨年龄之谜

该化石是产于一砾石层中,此层与上下地层关系并未完全搞清,且由于砾石层有些硅化而未能测其准确年代,而只能根据周围石笋和钙板的年代来推测。

4.Mystery of the Age of Nanjing Ape-man's No.2 Calvarium

The fossil was discovered in a gravel layer. The relationship between this layer and the layers up and down is in confusion. So his age can only be inferred by the ages of stalagmites and flowstone layer around, for there were some silicifications in gravel layer.

葫芦洞景观奇特
Unique Landscape of Hulu Cave

1、葫芦洞及猿人小洞

葫芦洞:平面呈东西走向,长65米,宽25米,面积约2000多平方米。天然洞口在洞的北边,洞东端游人出入口处为人工开凿。由于冲积物堆积将长形洞穴拦腰收束,而成葫芦状,故称葫芦洞。猿人小洞在洞南面地下,东西长8.26米,南北宽4.4米。

1.Hulu Cave and the Small Cave of Ape-man

Hulu Cave: it is in west-east direction, 65m long and 25m wide, more than 2,000m². The natural structural opening is in the north of the cave, and the visit opening in the east was dug artificially. Due to the alluvium deposited, the long cave became narrow in the middle, like a shape of bottle gourd, so given the name Hulu Cave. The small cave of ape-man is in the underground of the south of Hulu Cave, 8.26m long west-east, 4.4m wide north-south.

南京猿人(左图)与北京猿人(右图)鼻子的比较

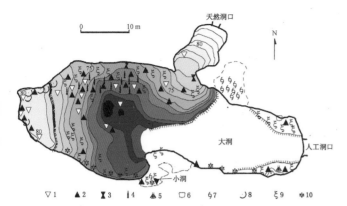

1 钟乳石；2 石笋；3 石柱；4 鹅管；5 石幔；
6 石幕；7 卷曲石；8 流石坝；9 钙板；10 石葡萄

南京汤山葫芦洞及小洞平面形态及化学沉积物的分布

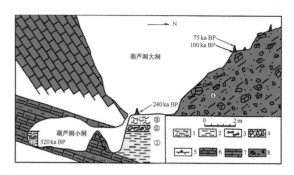

南京汤山葫芦洞洞穴堆积物剖面（汪永进等，1999）

1 斑杂状黏土层；2 黄红色黏土和泥质物，具纹层构造；
3 哺乳动物化石；4 钙质胶结角砾层；5 钙板和石笋；
6 杂色钙质砂泥岩；7 奥陶系灰岩；8 灰岩角砾岩
75ka, 100ka, 520ka, 240ka 分别代表根据石笋测定的年代

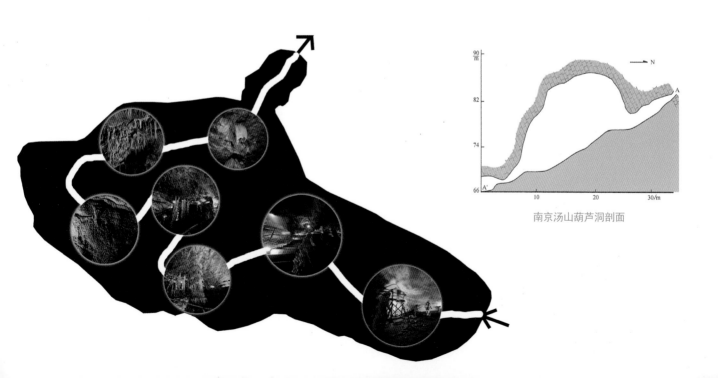

南京汤山葫芦洞剖面

2、溶洞怎么形成的？

发育溶洞的岩石是距今4亿多年奥陶系的石灰岩，其化学成分为碳酸钙（$CaCO_3$），属可溶性岩石。当石灰岩层处在地下水面以下，断层破碎带富含地下水。水从空气或土壤中吸收二氧化碳，形成了含碳酸的地下水，它对可溶性的石灰岩进行溶蚀。从小洞不断地扩大成大洞。后来地壳上升，地下水面下降后，洞内不再有水，就成了现今的葫芦洞。

在石灰岩岩溶（喀斯特）地貌的形成过程中，岩洞在接近或低于地下水位处形成。后来，地下水位下降，新的岩洞在低于地下水位处形成。岩洞塌陷或者岩洞表面溶解，在地表可形成落水洞。

2.How Was Karst Cave Formed?

Rocks developing karst cave were Ordovician limestones of more than 400 million years ago, and their chemical compositions were CaCO3, which belonged to soluble rocks. After lime rock layer was located below the water level underground, fault fracture zone got rich in groundwater. Water absorbed carbon dioxide from the air or soil, then became groundwater containing carbonate which corroded soluble rocks. Small cave became bigger and bigger. And then when the crust rose, the underground water level dropped, no water was in the cave, so Hulu Cave was formed.

In the forming process of Limestone karst landform, cave was formed near or below the groundwater. Later, the groundwater lowered down, new cave formed below the groundwater. Swallet hole can form in land surface when cave collapses or cave surface dissolves.

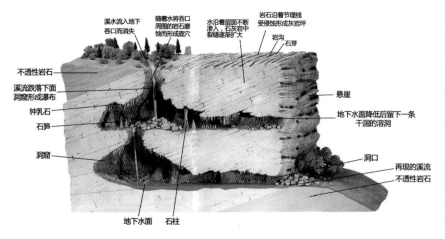

溶洞形成示意图

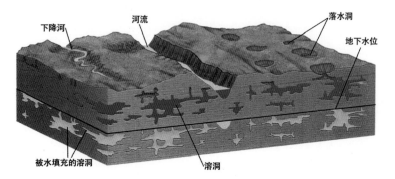

溶洞形成图解

3、溶洞的化学沉积物景观

进入溶洞可看到丰富多样的景观，钟乳石、石笋、石柱、鹅管、石幔、石葡萄、钙板。地下水中碳酸钙含量近于过饱和，当这些水往下滴时，由于蒸发失去了水分，水中的碳酸钙就会沉淀下来。

3. Karst Cave's Chemical Sediment Landscape

Entering a Karst cave, we can see rich varieties of landscape, such as stalactites, stalagmites, stone pillars, geese tubes, stone curtains, stone grapes, flowstones. The content of calcium carbonate in the groundwater is close to oversaturation. When water drops, calcium carbonate will precipitate due to evaporation of water.

知识链接
Knowledge Links

喀斯特

"Karst"原是斯洛文尼亚境内伊斯特里亚半岛（Istria Peninsula）上的一个地名，那里石灰岩广布，形成了一种独特的奇峰异洞地貌景观。

1893年南斯拉夫学者斯维奇（J.Cvijic）对该区地质地貌景观进行了较详细的研究。之后，"Karst"便成为对石灰岩地区溶蚀作用及其形成地貌的世界性通用术语。我国学者把"Karst"音译成"喀斯特"。

Karst

"Karst" was a place in the Istria Peninsula within the territory of Slovenia. There are limestones widespread, forming a unique landscape of peaks and different caves.

In 1893, J. Cvijic, a Yugoslavic scholar, carried out a more detailed study of this region's geological landscape. Then, "Karst" became a worldwide generic term which refers to the corrosion of the limestone areas and its formation of landforms. Generally, Chinese experts translate "Karst" to "喀斯特".

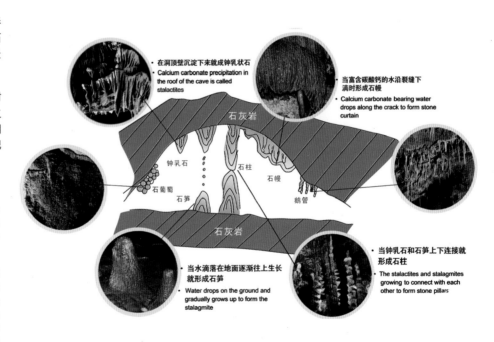

钟乳石形成示意图

4、水平钙板

在洞内水流如有堵塞而形成暂时性积水，积水中的碳酸钙淀积下来就成了碳酸钙沉积层。后来地壳抬升，昔日的沉积层就形成了洞中钙板。

4. Horizontal Flowstones

If there is water blockage in the cave forming a temporary accumulation of water, calcium carbonate will be deposited to be sedimentary formation of calcium carbonate. Later the land crust uplifts, the former sedimentary becomes flowstone in the cave.

5、洞穴泥石流堆积

大洞穴内北面自然洞口处有约20米厚的泥石流堆积层，它主要由土褐色砂质黏土、粉砂或砂砾杂乱堆积在一起。砾石多为灰岩石块，个别大的甚至有60~70厘米长，30~40厘米宽。泥石流之上有不规则的钙板层和石笋，经测定其年代约距今10万年，故10万年前正是这一巨大泥石流堆积物把洞口全堵着了，使洞内与洞外全隔绝了。

5. Accumulation of Debris Flow

There is a 20m-thick sedimentary formation of debris flow in the natural structural opening in the north of the cave, which includes tanglesome drab sandy clay, silt and gravel. Gravels are often limestone rocks, the individual large ones are even 60~70cm long, 30~40cm wide. There are flowstone layer and stalagmites of about 100,000 years ago above debris flow by determination. So about 100,000 years ago, it was the huge sedimentary formation of debris flow that blocked the cave and isolated inner cave from outside.

钙板

 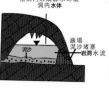 

水平层状钙板形成示意图

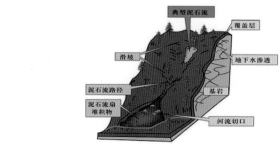

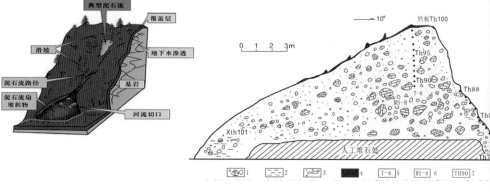

1、角砾层；2、黏土层；3、含哺乳动物化石层；4、石笋；5、热释光测试样品采样号；6、黏土结构采样号；7、系统采样号

南京汤山葫芦洞北边洞口泥石流堆积层示意图

阳山碑材
Yangshan Tablet Material 25

阳山碑材景区是汤山方山国家地质公园主景区之一，也是国家4A级旅游风景园区。阳山碑材是明孝陵大石碑工程遗址，也是与丰富地质遗迹相融合的地区。1996年第三十届世界地质大会在北京举行，会后来自美国、瑞典、南非等多个国家的专家考察后，异口同声地说："阳山碑材是文化与地质相结合的世界奇迹。"

阳山碑材园区有三大主题：古风遗韵的明文化村；守望600年的碑材；科普园地的石灰岩地层。

Yangshan Tablet Material is one of Tangshan-Fangshan National Geopark's main scenic spots. It is also a national 4A-class tourist scenic park. Yangshan Tablet Material is the large monument engineering site of Ming Xiaoling Mausoleum, which merges rich geological heritages. The 30th World Geological Congress was held in Beijing in 1996. Many experts from America, Sweden, South Africa, etc. investigated here after the congress. They said in unison, "Yangshan Tablet Material is a combination of cultural and geological wonders of the world."

Three Themes: Archaic Rhyme of Ming Villages; Tablet Material of 600 Years Ago; Limestone Formation of Popular Science Field.

明文化村、古风遗韵

园区入口处有阳山碑材牌坊，牌额正面刻书法家尉天池书写的"阳山碑材"四个镏金大字，背面镌刻原南京市委书记顾浩书写的"阳山碑材公园"六个大字。

明文化村是走进阳山碑材园区最先映入眼帘的一大片仿古建筑群，颇具明代文化气息的人文景点。建于2002年，房屋百余间，展示了一段传统的明代世俗画卷。如成记铁铺、张石匠屋、东厂督院、吉祥赌坊、金陵镖局、百姓书屋、御药房、南北杂货店、头饰店、工艺坊、游艺坊、甜趣坊、翠花豆腐坊、蓬莱酒坊、古泉问茶楼、关圣殿、古戏台、射箭场等。按当时的场景布置，店小二的吆喝、张铁匠的打铁声、织布坊忙织布、鞋店忙制鞋等，令人游览到此如身临其境，有回到明朝之体验。另外古戏台、大明湖上有文艺、武术表演，"皇帝祭碑仪式"、"重塑碑材"、"审逃犯"、"赐御婚"等广场剧，借助广场、市井街道和园区自然环境进行表演，游客们可体验明文化一个片段。

Ming Villages, Antique Life

There is a memorial arch of Yangshan Tablet Material in the entrance of the park. Four Chinese characters of "阳山碑材" were carved in the front written by the calligrapher Wei Tianchi. And six Chinese characters of "阳山碑材公园" were carved in the back written by Gu Hao, the former party secretary of Nanjing.

Ming Villages are antique buildings in the front of the park, full of cultural characteristics of Ming Dynasty. More than one hundred of buildings built in 2002 show a traditional secular picture of Ming Dynasty. Such as the Blacksmith Cheng's, the Stonemason Zhang's, Dongchang Office, Jixiang Casino, Jinling Escort, People Bookstore, Imperial Pharmacy, Grocery Store, Headdress Shop, Handicraft Shop, Play Shop, Sweets Shop, Cuihua Toufu Shop, Penglai Wine Shop, Guquan Tea House, Guan Yu Temple, Ancient Stage, Archery Range. Come here, you would like to go back to Ming Dynasty. There are many literature and art, martial arts performances on Ancient Stage and Daming Lake. Visitors can experience a fragment of Ming culture by watching the performance in the square, streets and natural environment of the park.

古井

位于"古泉问茶楼"前，古井已有600多年的历史，井深6米，曾于底泥中清理出十多只陶罐，井水属低钠、中等硬度的优质饮用泉水。

Ancient Well

The Ancient Well, 6 meters in depth, located in front of the "Ancient Spring Teahouse", has a history of over 600 years. Over 10 terrines have been cleaned out from the well sediment. The well water is high quality drinking water of low sodium and medium hardness.

袁机墓

袁机（1720—1759），字素文，浙江钱塘人，是清代文学家、江宁县令袁枚之妹。她与高绎祖指腹为婚。婚后受尽折磨，逃回娘家，常以诗文抒发悲伤的身世。乾隆二十四年因病去世，乾隆三十三年（1767）袁枚葬三妹素文于阳山，并写下凄楚动人的《祭妹文》，还为她编印了《素文女子遗稿》存世。

Yuan Ji Tomb

Yuan Ji(1720 — 1759), styled Suwen, a native of Qiantang, Zhejiang Province, is a sister of Yuan Mei who was the literati and the leader of Jiangning County of Qing Dynasty. She was pulp for the marriage with Gao Yizhu, from which she suffered enough and escaped back to her natal family. After that she often wrote poetries to express the sadness of life experience. She died of illness in the 24th of Qianlong. Yuan Mei buried her in Mt. Yangshan in the 33rd of Qianlong(1767), and wrote down tragically moving Offering Sacrifice to Sister and edited Suwen's Woman Posthumous Manuscript.

旷世碑材，史迹留痕

阳山碑材原址是阳山碑材园区最具历史感的主体景观，是省级文物保护单位。体验明文化村风貌之后，走进阳山碑材的遗址区。

Outstanding Massif, Historic Evidence

The former address of Yangshan Tablet Material is the most historic main landscape in the park of Yangshan Tablet Material and is the cultural relics under provincial level protection. Having experienced the Ming Cultural Villages, you'll step into the site area of Yangshan Tablet Material.

1、阳山碑材由来

阳山碑材是一个叫朱棣的皇帝为了给其父皇朱元璋建神功圣德碑而开凿的。

明代皇帝朱棣（1360—1424），在历史上做过几件大事：命郑和下西洋、令解缙主持修《永乐大典》、迁都北京、开通运河解决南粮北运。朱棣要为其父皇朱元璋建"大明孝陵神功圣德碑"，选在阳山开凿。在山体岩石上凿刻碑身、碑座、碑头。但未完工，留下了三块巨大的碑材，躺在阳山已有600多年，成为奇迹，令世人赞叹。

碑材开凿于明代永乐二年十月（1404年11月），至永乐三年八月（1405年9月）。征集劳工千余人，耗时300多天。

碑材工程终止时，朱棣命胡广等三学士考察阳山碑材。胡广的《游阳山记》中明确记载有"皇帝因建碑孝陵，断石于都城东北之阳山，其长十四丈有奇……色黝泽如漆……"。"仰见碑石穹然城立，山高数里，其体皆石。"《游阳山记》中还描述沿途山水田园，说"凡目之所见，耳之所闻，与夫一草一木之微，无不可乐。是皆圣天子之赐也"。

1. The Origin of Yangshan Massif

Yangshan Massif was cut off ordered by Yongle Emperor Zhu Di to build a magic and holy tablet in memory of his father Emperor Zhu Yuanzhang.

Yongle Emperor Zhu Di (1360–1424) did several great things in history, such as Zheng He's Voyages to the Western seas, Xie Jin's taking charge of compiling the Yongle Canon, moving the capital to Beijing, and opening the canal to solve the problem of South food shipped into the North. Zhu Di chose the Mt. Yangshan as the site to build the "magic and holy tablet" for his father. The body, pedestal and head of the tablet were chiseled on the Mt. Yangshan. But the project have not been finished and the three huge massifs have been left in the Mt Yangshan for over 600 years, which has been an amazing miracle.

The project proceeded from the October of the Yongle 2nd (November of 1404) to the August of Yongle 3rd (September of 1405) which took thousands of workers over 300 days.

When the project was terminated Zhu Di ordered three scholars including Hu Guang to inspect Yangshan Massif. Hu Guang recorded what he saw and heard in his travel notes–Travel in Yangshan.

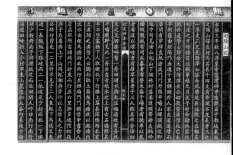

游阳山记

朱元璋　　　朱棣

胡广

碑身

2、碑材构成与体量——列入上海基尼斯之最

原地保存了碑材三大构件：碑座、碑身、碑头。

2. The Structure and Dimension of the Massif-Listed in the Shanghai Guinness Book of World Records

Three parts of the massifs were kept in Mt Yangshan: pedestal, body and head.

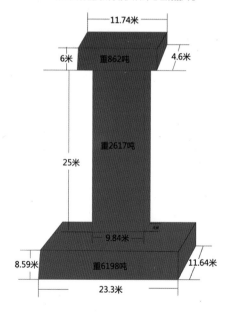

阳山碑材凿刻成孝陵大石碑毛坯规格尺寸

总高度39.59米
总重9677吨

碑身：其东北端尚与山体相连，其余三面已凿离山体。西北侧沿原岩层中一条断层，开凿成为宽1.2~3米堑沟，成为"一线天"。可以看出当时工程布置利用了原来岩石中断裂。凿开面上留下排列有序的凿痕，也留下了历史痕迹。碑身下部距地面约1.37米的上方凿出一个长27.6米，宽4米，高1.76米的矩形硐室，11个石墩。硐室以上碑材有效规格尺寸：高38.65米、宽10.52米、厚4米。若加工成大石碑毛坯尺寸：高25米、宽9.84米、厚4米。

碑头：四周已与山体凿离，底部距地面1.5~2.7米高的上方凿出高1.7米左右的六个硐室。碑材有效规格尺寸为高6米，宽18米，厚9.8米。若加工成大石碑毛坯，则高6米，宽11.74米，厚4.6米。

碑座：底部凿出斜长6.4米、斜宽7.6米、斜高1.78米的三个硐室。硐室上部碑材有效规格尺寸为高11.5米，宽16.1米，长23.3米。若加工成大石碑毛坯尺寸：高8.59米、宽11.64米、厚23.3米。

碑身、碑座、碑头三块碑材叠加总高度约56.15米，如最终凿刻加工成大石碑毛坯总高也有40米，毛坯总重量约为9,677吨。这巨大的碑材2005被年列为"大世界吉尼斯之最"。原为世界之最的宋代山东曲阜寿丘景灵宫前最大的单通碑材总高仅为20.33米，总重929吨，屈居老二。

Body of tablet：The northeastern part is still connected with the mountain while other three parts have been cut from it. A very narrow trench, 1.2~3m wide, is separated from the original rock stratum of the mountain in the northwest, forming "a thread of sky". It can be seen that the work was arranged to use the original rock fracture. The well-ordered chiseled traces were left on the cutting surface and so were the traces of history. A rectangular chamber(27.6 meters long, 4 meters wide, 1.76 meters high) and 11 stone piers were chiseled out on the lower part of the tablet body (1.37 m above the ground). The effective specification dimensions of the massif above the chamber are: 38.65 meters tall, 10.52 meters wide and 4 meters thick, which can be processed into big stone tablet blank in the specification: 25 meters tall, 9.84 meters wide, 4 meters thick.

The head of tablet: The tablet material was hewn from the mountain and six chambers about 1.7 meters tall were chiseled out at the bottom of the mountain which was 1.5~2.7m above the ground. The effective specification dimensions of the massif above the chamber are: 6 meters tall, 18 meters wide and 9.8 meters thick, which can be processed into big stone tablet blank in the specification: 6 meters tall, 11.74 meters wide, 4.6 meters thick.

The pedestal of tablet: Three stone chambers (oblique 6.4 meters long, oblique 7.6 meters wide, 1.78 meters tall) were hewn out on the bottom of the mountain. The effective specification dimensions of the massif above the chamber are: 11.5 meters tall, 16.1 meters wide and 23.3 meters long, which can be processed into big stone tablet blank in the specification: 8.59 meters tall, 11.64 meters wide, 23.3 meters thick.

The total height of three superimposed massifs is about 56.15 meters, which can be chiseled into a 40-meter stone tablet blank, the total weight of which is about 9,677 tons. The huge massif was listed in the "Shanghai Guinness Book of World Records" in 2005. The former highest massif had taken a back seat, which is in front of Shouqiu Jingling Palace of Song Dynasty in Qufu, Shangdong Province. The total height of the massif is only 20.33 meters, which weighs totally 929 tons.

山东（景灵宫）万人愁碑

3、选在阳山开凿碑材的缘由

为什么选在阳山开凿碑材？其一，这里采石已有悠久历史，考证始于六朝，古都南京的石刻、石碑，大多采自阳山；其二，岩石性质适宜，"色黝泽如漆"、"天赐良材"；其三，岩石中裂隙不发育，成材性能良好；其四，岩层的层面平缓易于凿刻成型。由此可见在600年前工匠已认识选材的一些地质知识。在开凿设计中考虑到碑材的边界利用岩石的断裂面，展示了古人的智慧。

3. The Reason to Choose Mt Yangshan to Chisel Massifs

Why do people choose Mt Yangshan to chisel the massifs? First, quarrying here has a long history which dates back to Six Dynasties. Most of the stone carvings and stone tablets of Nanjing are quarried from Mt Yangshan. Second, the rocks' properties are suitable and the color is as black as paint. Third, the fissuring in the rock doesn't develop and the rocks are suitable to become tablet materials. Fourth, the rock stratum is gentle enough to chisel. Thus, it seems that the craftsmen 600 years ago realized some geological knowledge on selecting materials. They considered using the rock surface fracture as the borders of massifs during the cutting design, which showed the wisdom of the ancients.

4、袁枚与《洪武大石碑歌》

观赏阳山碑材，不能不提到袁枚与《洪武大石碑歌》。袁枚（1716—1797）清代著名诗人，乾隆进士，曾任江宁知县。袁枚曾到阳山观赏碑材，并写了《洪武大石碑歌》：

青龙山前石一方，弓尺量之十丈长。
两头未截空中央，旁有赑屃形更大。
直斩奇峰为一座，欲负不负身尚卧。
高皇开创气概雄，欲移此碑陵寝中。
大书功德告祖宗，压倒唐汉惊羲农。
碑如长剑青天倚，十万骆驼拉不起。
诏书切责下欧刀，工匠虞衡井中死。
芟刈群雄答八荒，一拳顽石敢如此！
周颠仙人大笑来，天威到此几穷哉。
但赦青山留太仆，胜扶赤子上春台。
丁丁从此仃开凿，夜深无复山灵哭。
牧竖宵眠五十牛，此碑千载空悠悠。
昭陵石马无能战，汉代铜仙泪不流。
吁嗟乎！君不见项王，拔始皇鞭。
山石何赏不可迁，威风一过如轻烟。
惟有茅茨土阶三五尺，至今神功圣德高于天。

诗中描述了碑材的位置、体量。诗中有一句"碑如长剑青天倚，十万骆驼拉不起"，也写出劳工完不成考核工作量而被害，被迫投井的惨状。

4. Yuan Mei and Hongwu Large Stone Tablet Song

While enjoying the sight of Yangshan Massif, we can't fail to mention Yuan Mei and Hongwu Large Stone Tablet Song. Yuan Mei (1716 –1797), Qianlong Jinshi(a successful candidate in the highest imperial examinations), is a famous poet of Qing Dynasty who once was the leader of Jiangning County. Yuan Mei once came to Mt Yangshan to view massifs and wrote down Hongwu Large Stone Tablet Song. The poet describes the location, size and weight of the massifs and also writes the miserable situation of those workers who were killed or forced to drown themselves into wells because of not fulfilling the assessment workload.

大明孝陵神功圣德碑

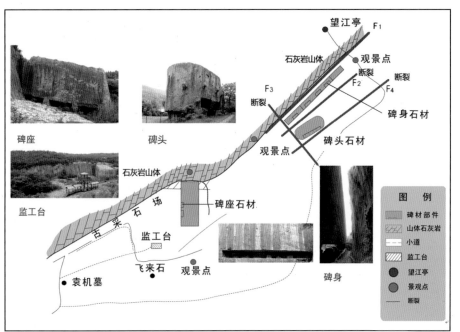

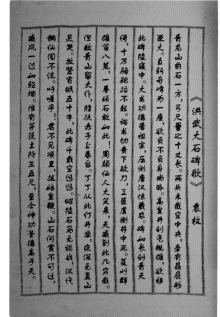

5、碑材开凿劳工与坟头村

据《大明会典工部·山陵》规定，当时开凿碑材有设计施工的官员与监工，设有监工台。劳工有近千人，其中有囚徒。袁枚诗中也说当时劳工凿碑之难、之惨。阳山碑材山脚下不断出现的一个一个新的坟头就是当时劳工惨死安息之地。固此其附近的村庄被称为坟头村。

5. Labor Engaged in the Massif-chiseling and Grave Mound Village

According to the provisions of Code of Great Ming Dynasty • Ministry of Works • Mountain , the project of massif-chiseling had officials who were responsible for design, construction and supervising, and supervision stations. There were thousands of workers including some prisoners. Yuan Mei described in his poem how miserably and hard they worked. New grave mounds which continually appeared at the foot of Mt Yangshan were the resting places for those workers with tragic death. Hence the nearby village was called Grave Mound Village.

6、阳山碑材没有运走建碑之谜

阳山碑材所在地是600多年前皇家开凿孝陵碑材的原址。大石碑已成雏形，但最终被朱棣放弃了。这成了阳山碑材的一个谜题。分析原因可能是多方面的：第一，工程未完工、石碑部分仍与山体相连，要全部完工的工程量仍然很大；第二，或许最重要的原因，是碑材体量太大，运输是个大问题。山东景灵宫碑材重929吨，用了15年时间运到目的地，有"日挪卧牛之地"之说。正如民谣说"东流到西流，锁石锁坟头；东也流，西也流，神仙也摇头；若要碑搬家，除非山能走。"当然还可能有其他原因，留待人们去思索、考证。

袁枚说：十万骆驼拉不起，民间说若要碑搬家，除非山能走。600年前的碑材的故事，也引起了当代青年人的思考，田家炳中学的几位学生想出了用轨道与滑轮运输的方法。

6. The Mystery of Yangshan Massif Failing to Build Tablets

The location of Yangshan Massif is the former address of massifs of Tomb of Emperor Zhu Yuanzhang. The large stone tablet had already taken shape but eventually been abandoned by Zhu Di, which is a mystery of Yangshan Massif. The possible reasons are various. First, the project was not finished and part of tablet was still connected with the mountain. Large quantities of labor force were still needed to complete the project. Second, maybe it is the most important reason that the massif is too large to carry away. The massif of Jingling Palace in Shandong weighs 929 tons which took workers 15 years to carry to the destination. It once was said that it could be moved a distance of a cow per day. Of course, there may be other reasons which are left to later generations to think and research.

Yuan Mei said that one hundred thousand camels couldn't pull them and folk said that the massifs couldn't move away unless the mountain could walk. The story of massif 600 years ago aroused the contemporary young people's thinking. Several students of Tian Jiabing High School came up with a transportation method using track and pulley.

石灰岩层、科普园地

阳山碑材不仅仅是历史遗迹，阳山碑材的岩石以及矿物、化石等具有神秘的科学故事。

1、石灰岩层

阳山开凿建碑的石头，叫石灰岩。石灰岩是海洋环境沉积形成的碳酸盐岩，以碳酸钙（$CaCO_3$）成分为主，常混有黏土、粉砂等杂质。岩石呈灰或灰白色，若混入碳质则呈灰黑或深黑色，性脆，硬度不大，小刀能刻动，滴稀盐酸会强烈起泡。

阳山古采石场的灰岩层地质年代为距今2.8亿年前的二叠纪早期，称为栖霞灰岩。栖霞灰岩自下而上分四个岩性段：臭灰岩层、下硅质层、含燧石灰岩层、上硅质层。开凿碑材的岩石选在燧石灰岩层段中，质较纯的灰黑色厚层石灰岩，其下部为含燧石灰岩。

Limestone Formation, Popular Science Field

The Yangshan Massif is not only historical remains but also has some mysterious stories about its rocks, minerals and fossils.

1. Layers of Limestones

The rocks which were used to chisel and build tablet are called limestones. Limestone is a carbonate rock, formed by the deposition of the marine environment which consists mainly of calcium carbonate ($CaCO_3$), often mixed with clay, silt and other impurities. Gray or off-white rock, mixed with carbonaceous turning dark gray or dark, is brittle with low hardness, which can be engraved with the knives. If dilute hydrochloric acid is instilled into the limestone, it will be strongly blistering.

The geological age of limestone layer in Yangshan ancient quarry is early Permian about 280 million years ago which is called Qixia Limestone. Qixia Limestone can be divided into four lithologic sections from the bottom to top : smelly limestone layer, lower siliceous layer, limestone layer containing flint, upper siliceous layer. The gray and black heavy bedded limestone in the flint limestone layer section is chosen to chisel the massif, the lower part of which is limestone containing flint.

石灰岩在海洋中形成

珊瑚礁石　燧石灰岩　珊瑚石灰岩　珊瑚碎屑石灰岩　藻类石灰岩

2、燧石

在石灰岩中常会见到黑色团块状硅质物称燧石，俗称"打火石"，又称燧石岩。它是一种致密坚硬的硅质岩，主要成分为微晶或隐晶石英组成的岩石。石灰岩在成岩过程中硅质（SiO_2）交代碳酸钙而形成的。燧石大多为团块结核状，直径大小不等，从几厘米至三四十厘米，还有呈扁头状、串珠状透镜体等。

2. Flint

A black lumpy siliceous material is often seen in the limestone which is called flint, commonly known as "Flintstones", also known as flint rock. It is a kind of dense, hard siliceous rock which mainly consists of microcrystalline or implicit quartz rock. During the formation of limestone, the metasomatism of siliceous material and calcium carbonate formed clumpy and nodular flint with diameters ranging from several centimeters to 30~40 centimeters. There are also flints in other forms, such as flat-headed shape and beaded lens.

燧石

3、方解石脉

在古采石场南段的黑色灰岩裂隙内，时常见到一些白色矿物，呈条带状、树枝状、网络状，其成分也是碳酸钙（$CaCO_3$），矿物名称叫方解石，它充填在灰岩裂隙内，称为方解石脉。

3. Calcite Vein

In the southern section of ancient quarry, it can often be seen that some white minerals distributed in banded, dendritic, network-like shape are in the fissures of black limestone, the ingredient of which is calcium carbonate ($CaCO_3$), the mineral name of which is calcite. It filled the fissures in the limestone which is called calcite vein.

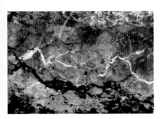

方解石网状脉（白色）

4、文石与钟乳石

碑身西南侧堑沟向西延伸方向，在未经凿刻的山体岩壁断层面上，有一小片棕黄色、灰白色文石、钟乳石。其成分都是碳酸钙（$CaCO_3$），前者呈小柱状、针状、放射状等晶形，后者呈小葡萄状、鲕状。

4. Aragonite and Stalactite

In the direction of the trench to the west on the southwestern side of the tablet there is a small area of brown-yellow, off-white aragonite and stalactite both consisting of calcium carbonate. The former is in the small columnar, acicular, radial crystal form and the latter is in the grape-like or oolitic form.

5、化石

古生物化石，是指人类史前地质历史时期形成并赋存于地层中的生物遗体和活动遗体化石及其遗迹化石。

在碑头、碑身、观景步道等处可见珊瑚、腕足类、腹足类和头足类鹦鹉螺等化石。

珊瑚化石：珊瑚为底栖的海生动物，有单体和复体。在碑头石材东南侧中部硐室下方，见到两个近似圆形的珊瑚化石。古采石场和碑身上也可见到珊瑚化石。

腹足类化石：腹足类动物属于软体动物门腹足纲（俗称螺虫帅）。它们广泛分布在海洋里、淡水中和陆地上。软体分为头部、足部和内脏区三部分。体外常具有一个螺旋形、圆锥形、笠形或平旋形的螺壳。在碑身石材上、碑座南侧山坡的观景步道上都有腹足类化石分布。

四射珊瑚化石

5. Fossil

Fossils refer to the biological remains and activities relics which are formed in the prehistoric geological history period and deposited in the stratum.

The fossils of corals, brachiopods, gastropods and cephalopods Nautilus can be seen in the head, body of tablet and scenic trails, etc.

Coral fossils: Corals are benthic marine animals including monomer and complex animals. Under the chamber in the southeastern side of the tablet head stone, there are two roughly circular coral fossils. Ancient quarries and the tablet body also have coral fossils.

Gastropod fossils: Gastropod animals belong to Mollusca Gastropoda (commonly known as snail insects handsome). They are widely distributed in the oceans, fresh water and on land. The soft part is divided into three parts of the head, foot, and visceral areas. The outside often has spiral-shaped, conical, donning shaped or planipiral form shells. The gastropod fossils are distributed on the stone materials of tablet body and the scenic trails which are in the south hillside of tablet base.

珊瑚　　　　腕足类　　　　腹足类　　　　鹦鹉螺

头足类鹦鹉螺化石：属于软体动物门。生活在海水里，头部有触手，用以捕食、爬行或游泳，故名头足。化石仅保存外壳，壳直、弯曲或松卷。壳体的形态与特点是研究鹦鹉螺的重要依据。在碑头西部硐室下方岩壁上能找到个体较小的鹦鹉螺化石。

Cephalopods Nautilus fossils: They belong to Mollusks. They lived in the seas, with tentacles on their heads for prey, crawling or swimming, hence the name, cephalopods. The fossils only kept the shell, the shell straight, bent or loose volume. The shape and characteristics of the shell are the important basis for researching Nautilus. Under the chamber on the western side of tablet head, small Nautilus fossils can be found on the stones.

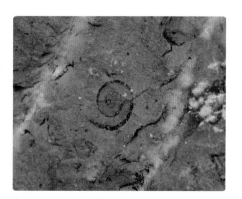

头足类鹦鹉螺化石

化石是怎样形成的？

生物死亡以后（包括遗体、遗迹或遗物），被沉积物埋藏起来，后被矿物质充填或交代，就形成化石。

How Fossils Are Formed?

When the live beings (including the remains, ruins and relics) died, they were buried by sediments, then filled or replaced by mineral materials, and at last formed fossils.

保护古生物化石的科学意义

古生物化石是研究进行地球演变、生物进化等的重要资料，是确定地层时代的重要依据。这是不可再生、不可移植的地质年代与生物生存的天然记录。这里化石只能欣赏观察，可以拍照观赏，切勿敲打。让我们用心去保存这份重要的地质遗迹。

The Scientific Significance of the Protection of Fossils

Fossils are the important materials to study the Earth's evolution and biological evolution. It is also an important basis to determine the age of the strata. It is a non-renewable, non-portable natural record of geological age and biological survival. Fossils here can only be appreciated and observed, photographed, but not knocked. Let us try our best to protect these important geological relics.

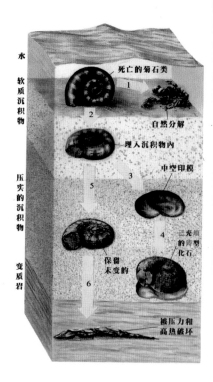

化石形成过程

①螃蟹在泥砂上生活着。

②螃蟹死亡后沉躺在泥砂上面。

③泥砂不断沉积，覆盖在螃蟹遗体上面，将它埋入泥砂之中。

④螃蟹的肉体渐渐消失，留下不易腐化的硬壳。

⑤硬壳有的被完整保留下来，石化成化石。有的则全部被溶蚀不见，只留下硬壳的模子。

双枚贝化石

螃蟹化石

6、断裂、断层与节理

断裂：是岩石形成后，在地壳变动的过程中受到外力的作用发生破裂的现象。

断层：是在岩层发生破裂时，由于受到不同方向力的差异，使其两侧的岩层发生相对的移动，可分为正断层、逆断层、平移断层。

节理：是岩层发生破裂产生的较小的裂缝，是没有明显位移的断裂。

6. Fractures, Faults and Joints

Fractures are rupture phenomena appearing at the time when the rocks are exerted by the action of external forces during the process of changes of the earth's crust after their formation.

Faults are the relative movements of rocks of both sides occurring at the time when the rock ruptures due to the difference of different direction forces. Faults can be divided into the normal faults, reverse faults, slip faults.

Joints are small cracks generating from the ruptures of the rock stratum. They are the fractures without significant displacement.

断层类型 Fault type

断层产生前 Before fault produced

正断层 Normal fault

逆断层 Reverse fault

平移断层 Strike-slip fault

阳山碑材园区也见到断裂

1号断层（F1）：在碑身的西北侧，规模较大，走向50°~230°，倾向南东，倾角近似直立（85°~90°）。由灰岩角砾形成的破碎带宽1-3米，明代石匠沿此处凿刻碑身而成为堑沟，形成一线天景观。西北端未开凿的断层角砾都清晰可见。这说明当时的工匠已利用这一条断裂作为碑身的边界。

3号断层（F3）：位于碑身和碑头西南端的外侧，断层走向145°~325°，倾向北东，倾角约70°，破碎带宽3~5米。

5号断层（F5）：位于碑座西侧约60米处的山体陡壁上。厚约3米的薄层状石灰岩与薄层状碳质页岩呈互层的岩层在断开时发生位移，左下右上移动距离约3米。F5断层走向330°，倾角向北东，倾角约80°，为逆断层。（参见第32页图）

Fractures in the Yangshan Massif Park

Fracture 1(F1): It is on the northwestern side of the tablet on a large scale which moves toward southeast in the angle of 50°~230°. The inclination is approximately vertical (85°~90°). The width of crushed zone formed by limestone breccia is 1-3 meters, along the craftsmen of Ming Dynasty chiseled a trench which formed the scene of "a thread of sky". The not-chiseled fault breccia on the northwestern end can be seen clearly which indicates that the craftsmen used the fracture as the border of the tablet.

Fracture 3(F 3): It is located on the outside of the southwestern end of the body and head of tablet which moves toward northwest in the angle of 145°~325°. The inclination is about 70°. The width of the fracture is 3~5 meters.

Fracture 5(F5): It is located on the steep mountain cliff which is 60meters away from the western side of tablet base. The 3-meter thick laminated limestones and laminated carbonaceous shales are the alternating layers of rock, between which the displacement took place. The moving distance is about 3 meters. The F5 moves toward northwest in the angle of 330°, the inclination of which is about 80°. It is a reverse fault.

阳山早二叠世栖霞灰岩及小断层

汤山温泉
Tangshan Hot-spring
39

汤山温泉概况

分布：汤山7眼温泉分布在汤山山体东部的山脚下，沿汤水河西岸地面绕山分布，泉眼出露地面标高自39.40米至43.43米。

温度：50~59℃。

品质：汤山温泉水中含有钾、钠、钙、镁、铜、锌、锰等对人体有益的矿物质、微量元素。温泉水质为SO_4~Ca型，pH值6.7~7.4，矿化度1.59~1.85g/L，水中氟含量3.2~5.0g/L，偏硅酸含量50-62mg/L，达到医疗矿水命名的浓度标准，属弱碱性硫酸钙型氟、偏硅酸医疗热矿水。锶含量5.23~7.58mg/L，接近命名浓度。

The General Situation of Tangshan Hot-spring

Distribution: The Tangshan hot-spring with seven outlets where the spring has been gushing is distributed in the east foot of Tangshan Mountain, surrounding the mountain along the west bank of Tangshan river, with the seven outlets from 39.40m to 43.43m above the ground.

Temperature: 50~59℃.

Quality: Tangshan hot-spring is abundant in minerals and microelements, including potassium, sodium, calcium, magnesium, copper, zinc and manganese, which are beneficial to the health.

Water quality: SO_4~Ca. pH:6.7~7.4. The degree of mineralization; 1.59~1.85g/L. Fluorine in the water: 3.2~5.0g/L. Metasilicic acid in the water: 50~62mg/L. Strontium in the water:5.23~7.58mg/L. The water in Tangshan hot-spring reaches the concentration standards of mineral water with medical value.

1、什么是泉与温泉

地下水从岩石或土壤中自行流出地面来，称为泉。按泉水流出的方式可分为上升泉：由于水的静压力或所含气体的推动而涌出地面；下降泉：借重力作用自高处向低处流出。我国一般以年平均气温20℃为标准，若泉水温度≤20℃称为冷泉；≥20℃-≤37℃为温泉；≥37℃-<42℃为热泉；≥42℃-99℃为高温泉；≥100℃为沸泉。云南省红河边上的一眼温泉喷出时水温达103℃，是温度最高的泉水。汤山温泉水温50~59℃，称为高温泉。

1.What Is Spring and Hot-spring?

Spring is the groundwater flowing to the surface of the Earth from the rock stratum and soil naturally. Springs are classified as ascending spring and descending spring according to the way the spring flows. Ascending spring is pouring from the underground by static pressure of water and the push from the gas in the water; descending spring is flowing from a high place to a low place with gravity. Generally our country develops a spring classification based on the average annual temperature of 20℃. If the average annual temperature is below 20℃, it is called cold spring; if the temperature is between 20℃ and 37℃, it is warm spring; if the temperature is between 37℃ and 42℃, it is hot spring; if the temperature is between 42℃ and 99℃, it is hyperthermia spring; if the temperature is above 100℃, it is boiling spring. The hot-spring with one outlet along the Hong River in Yunnan Province is the hottest spring, the temperature of water from it reaching 103℃. The temperature of Tangshan hot-spring is between 50~59℃, and therefore it is called hyperthermia spring.

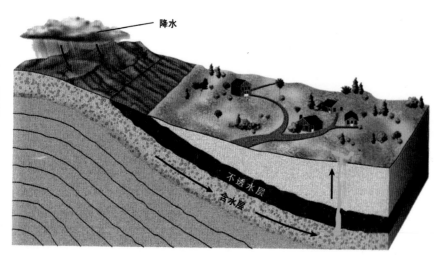

温泉形成——雨水，地下水

温泉含矿物质多的叫矿泉，根据主要矿物成分不同，分为硷泉、盐泉、铁质泉、苦泉、硫黄泉等。

按照泉水出现的地质环境，又可分为接触泉、断层泉、裂隙泉等。

我国的温泉地热资源十分丰富，已探明有2 600多处。江苏有10余处，汤山温泉地热资源，其水温、水质、水量列为全国十大知名温泉。

The hot-spring containing many minerals are called mineral springs. According to the major chemical composition of the water produced, the hot-springs can be classified as alkali springs, salt springs, chalybeate springs, bitter springs, sulfur springs and so on.

In accordance with the geological environment where the hot-springs gush, it also can be classified as contact springs, fault springs, fracture springs and etc.

Our country has an abundance of geothermal resource, having discovered over 2 600 hot-springs. Jiangsu has more than 10 hot-springs and Tangshan hot-spring is listed as one of National Top Ten Famous Hot-springs with its pleasant temperature, high water quality and volumes of water.

2、汤山温泉的形成

· 汤山地下水储存在哪里？

组成汤山的主要岩层为距今4.4亿年前的奥陶系石灰岩，总厚度近3 000米，岩层陡立，层理明显，经多期多次构造变动（造山运动），断层、裂隙、溶洞发育，地表水进入地下后，连通性好，构成良好的地下水运行网络，成为地下水的储水库。

2.The Formation of Tangshan Hot-spring

·Where is the Tangshan groundwater stored?

The major rock stratum which constitutes the Tangshan is Ordovician limestone 440 million years ago and about 3,000m wide.

The steep rock stratum with clear texture has experienced much orogeny: faultage, crack and the forming of cave and then the surface water flowing into the underground. Thus a good network of groundwater was woven, turning into the reservoir of groundwater.

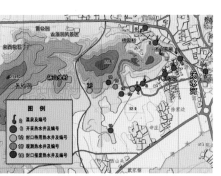

汤山东区地热井分布图

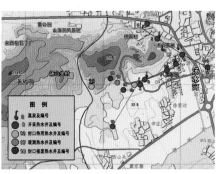

汤山西区热水井分布图

•温泉水的来源

主要来自天水，当大气降水（落雨）到达地面渗入地下，其余汇流变为流水，亦有一部分渗入地下，渗透水是地下水的主要来源。

•地下水为什么是热的？

热源主要来自两个方面：一是地热增温，增温率为3%，即每深入地下100米，地温就增高3℃，地下水温也提高3℃。这种热能是地壳内部放射性元素蜕变而产生的。二是与岩浆活动和断裂构造有关。在距今约1.2亿年前岩浆沿断层裂隙侵入到汤山地区的石灰岩层，形成岩体、岩脉，可能有残余热能。由这两种热能使地下水成为热水。

•温泉形成还要有盖层

盖层：距今约4.2亿年前的志留系（S）至距今约1亿年前的白垩系下统（K），厚约3000余米的一套砂页岩，泥质岩地层，透水性差、裂隙不发育，分布在热储层部分地段的上部或周边，为良好的隔热盖层。

•地下水热水是怎样上升形成温泉的？

汤山温泉在成因上属深层循环地热增温，兼与岩浆活动有关的地热流体。断裂构造成为深部地热流体向浅部运移通道，成为地下热水溢出形成汤山温泉的重要条件之一。当地下热水沿着纵横交错的破碎带和溶洞，由高向低流动，遇到不透水的火成岩—石英闪长斑岩或砂页岩阻挡，地下水就沿断层，裂隙上升涌出地面，形成温泉。

•The resource of hot-spring

The hot-springs mainly come from meteoric water. When it rains, some of the water flows together into running water and the other infiltrates into the underground, which is a main major source of groundwater.

•Why is the underground water hot?

There are two sources of heat that heat the groundwater. One is geothermal heat. The rate of temperature increase with depth is 3%, that is, the 100m deeper, the 3℃ higher the temperature of the groundwater is. The heat is sent out from the spallation of radioactive elements in the crust. The other is related to the movement of magma and cracks and fissures. Magma infiltrated into the limestone stratum in Tangshan area along the cracks and fissures about 120 million years ago and formed rock mass, dyke and left some thermal energy.

•Caprock is a necessity in the formation of hot-springs.

Sandy shale in Tangshan area, formed between the Silurian about 420 million years ago and the Cretaceous about 100 million years ago, over 3000m deep, is covering or surrounding the part of heat stratum and is the perfect caprock that insulates heat due to the pelite quality, poor water permeability and undevelopment of cracks.

•How does the hot groundwater emerge as hot-springs?

Tangshan hot-spring is heated by the cycling geotherm in the deeper stratum together with geothermal fluid connected with the movement of magma. A network of cracks and fissures has been the passageway for the flow of geothermal fluid from deeper part to the shallower part and is essential in formation of Tangshan hot-spring. When the hot groundwater comes across the staunch petrosilex or is resisted by sandy shale in the flow from the high to the low along the criss-cross network of cracks and caves, groundwater will issue to the surface of Earth along fault and cracks; therefore, the hot-spring is formed.

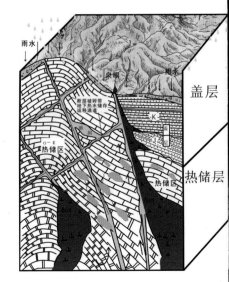

汤山温泉热水形成机理示意图

汤山温泉文化

1、刘义恭题诗《汤泉铭》

南朝宋，江夏王刘义恭（413—465），宋武帝之子。他是1,500多年前为汤山温泉写诗第一人。相传，当年这位"善骑马、解音律、游行或三五百里"的王子来到汤山一带游览。他策马蹚过汤水河来到汤山脚下，见热气腾腾的温泉水从山中涌出，十分高兴，就有感而发，作《汤泉铭》诗一首，将汤山温泉与秦都咸阳温泉、汉京长安骊山温泉媲美。

     汤泉铭
   秦都壮温谷，汉京丽汤泉。
   炎德资远液，喧波起斯源。

2、袁枚情系汤山温泉

袁枚（1716—1798），乾隆进士，曾任江宁等地知县。他所作组诗《浴汤山五绝句寄香亭兼谢荷明府》，深情地表达了对汤山温泉的热爱。

  为寻圣水濯尘缨，爱忍春寒远出城。
  刚是杏花村落好，牧音相约过清明。
  方池有水是谁烧？暖气腾腾类涌潮。
  五日熏蒸三日浴，鬓霜一点不会消。
  延祥寺里证前因，二十年前借住身。
  今日僧亡菩萨在，应知我是再来人。
  野外闲行乐有余，阿连底事劝回车。
  天生此水温存性，只恐妻孥转不如。

The Culture of Tangshan Hot-spring

1. Liu Yigong and Tangshan Hot-spring Inscription

Liu Yigong(413–465), Jiangxia King in the Song Kingdom of the North and South Dynasties, is the son of Liu Yv, Emperor Wu. He was the first poet who wrote for Tangshan hot-spring more than 1500 years ago. It is said the prince who was good at riding a horse, proficient in music and covered thousands of kilometers came to visit Tangshan area. He spurred across the Tangshui River and got to the foot of Tangshan Mountain. A stream of hot spring water gushing from the mountain came into his sight. He was so excited that he wrote Tangshan Hot-spring Inscription and compared Tangshan hot-spring with Xianyang hot-spring in the capital of Qin Dynasty and Lishan hot-spring in the capital of Han Dynasty.

2. Yuan Mei and Tangshan Hot-spring

Yuan Mei(1716–1798), a successful candidate in the highest imperial examination during the reign of Emperor QianLong, served as the magistrate of Jiangning county and other places. The poet deeply expressed his love to Tangshan hot-spring.

3、汤山蒋介石温泉别墅

该别墅坐落于温泉路3号，原先是国民党元老张静江于1920年建造的花园式温泉别墅，原名"张公馆"。主楼坐北朝南，分上下两层，一楼下半层嵌于地下仅上半层露出地面，泉水经管道直接流到别墅的浴池。

1927年12月1日，蒋介石与宋美龄结婚，张静江将这处温泉别墅作为贺礼赠送给蒋介石夫妇，12月9日到此入住。其后就经常来此居住，办公，接待宾客等。

3. Chiang Kai-shek's Villa with Spa in Tangshan Mountain

The villa, located in No.3 Hot-spring Road, was originally built as the garden-like villa with spa by Zhang Jingjiang, the senior statesman of the Kuomintang, in 1920 and it was first named "Mansion for Mr. Zhang". The main building faces south and has two floors. The lower part of the first floor is embedded in the earth and the upper part of it is above the ground. The spring flows directly into the bathing pool through pipelines.

Mr. Zhang presented the villa to Mr. and Mrs. Chiang Kai-shek when they got married on Dec.1st, 1927. They came to live here on Dec 9th, 1927 and after that they often lived, worked and entertained guests here.

（2002年公布为省级文物保护单位）

(listed as one of the provincial cultural relics protection units in 2002)

4、陶庐温泉别墅

陶庐位于温泉路1号院内。江宁士绅陶保晋（1875—1948）于民国八年（1919）率先在汤山建温泉别墅，一座两层楼13间房屋中有浴池6个，两栋平房，一栋有房5间，男浴池3个，另一栋有房1间，女浴池2个。

陶庐初期只供家人和亲朋好友使用，后期名气大增方公开对外营业。

4.Taolu Villa with Spa

Taolu Villa is situated in No. 1 Hot-spring Road. Tao Baojin(1875–1948), a gentry from Jiangning, took the first in building villas in Tangshan in 1919. The villa consists of one two-storey building with 6 bathing pools and two bungalows, one of which has 5 rooms and 3 bathing pools for men and the other has one room and 2 bathing pools for women.

At first the bathing pools in Taolu Villa were only for his family, relatives and friends. Later on, these bathing pools were open to the public with its increasing popularity.

5、圣汤延祥寺

韩滉（723—787），唐京兆长安（今陕西西安）人。唐德宗时，任浙西观察使，是一位政治家、画家。他的小女有恶疾，到汤山洗温泉浴治愈小女顽固的皮肤病后，韩滉不惜拿出女儿的嫁妆钱，在温泉畔建造了一座寺院，以谢"圣汤"。寺院名就叫圣汤延祥寺。宋代王安石，清代袁枚等先后来到圣汤延祥寺题诗颂扬。寺院建筑于1970年前后被拆除。

5. Yanxiang Temple in Tangshan

Han Huang(723–787), born in chang'an (Xi'an in Shaanxi Province) in Tang Dynasty, was a statesman and painter and once served as an official in charge of military and political affairs in western part of Zhejiang during the reign of Emperor Dezong of Tang Dynasty. His little daughter was seriously ill and went to bathe in the Tangshan hot-spring. After that, the stubborn skin disease of his little daughter was cured. In order to appreciate the "sacred Tangshan hot-spring", he spent the money which had been originally set aside as the daughter's dowry in building a temple named Yanxiang Temple. Wang Anshi in Song Dynasty, Yuan Mei in Qing Dynasty and many other poets came to the temple to inscribe poems to extol the Tangshan hot-spring one after another. The temple was torn down around 1970.

6、江苏省工人汤山疗养院

汤山疗养院为江苏省总工会直属大型疗养院。始建于1956年，现有客房大楼5幢，床位500余张。温泉水直通客房。治疗中心设有室内游泳馆、器械水疗、水按摩、温泉桑拿、健身房、娱乐室等。在皮肤病、骨关节疾病的治疗，心血管疾病康复等方面有独到之处。

6. Jiangsu Tangshan Sanatorium for Workers

Tangshan Sanatorium, built in 1956, is affiliated to Jiangsu Provincial Federation of Trade Unions. It has 5 buildings of guest rooms, with more than 500 beds. The hot-spring water is avraiable in every guest room. The therapeutic centre is equipped with a natatorium, a spa room, a hydromassage room, a hot sauna room, a gym, and a recreation room, which is especially useful in cure of skin diseases and osteoarticular diseases and rehabilitation of cardiovascular disease.

7、南京军区汤山疗养分院

该疗养院位于汤山镇温泉路5号，院内有3个温泉。1932年时任考试院院长的戴季陶在汤王庙附近建别墅，1937年被日军炸毁。12月15日下午日军占领汤山，看中了这里的温泉。次年就在此建造规模较大的伤兵医院。抗战胜利后，国民政府将其改建为联勤医院。解放后成为解放军八三医院。如今已建成具有特色的温泉疗养院。

7. Tangshan Sanatorium of Nanjing Military Region

The sanatorium, located in No. 5 Hot-spring Road in Tangshan Town, has 3 hot-springs. Dai Jitao, the president of Examination Authority in 1932, built a villa near Tangwang Temple, which was blown up by Japanese troops in 1937.Japanese troops occupied Tangshan Town on the afternoon of Dec. 15th, took a fancy to the hot-spring and built a large hospital for the wounded soldiers the next year. After the victory of the Anti-Japanese the War, national government rebuilt it to a hospital affiliated to Joint Service Department and it became Eight-Three Hospital of PLA after liberation. Nowadays it has become a spa sanatorium with some specialties.

江苏省工人汤山疗养院

南京军区汤山疗养分院

名人与温泉有关的活动片段
The Notables' Activities Related with Tangshan Hot-spring

1、张静江

张静江夫妇　　1927年4月，民国政要在张静江的汤山温泉别墅沐浴后合影。

3、于佑任

于佑任，在汤山宁杭公路旁的黄墅村建造了一座温泉别墅，有屋10余间，内设温泉浴室。

2、宋美龄

宋美龄沐温泉浴后散步向蒋介石提出拨款修扩建小学校舍，蒋当即应允拨款，1928年建成投入使用。主建筑于1937年12月被侵华日军焚毁。

4、张学良

张学良曾三次到过汤山，沐温泉浴，每次在南京逗留均为一个月，蒋介石每次都陪他到汤山的温泉别墅沐浴。

旅游咨讯
Tourist Information | 47

主要景区

1、绝世碑材——阳山碑材（明文化村）

阳山碑材（明文化村）景区是国家4A级旅游区，省级文物保护单位，由演绎打造碑材场景的明文化村、阳山石灰岩层和世界之最的阳山碑材三部分游览区域组成。

特色餐饮：阳山香豆腐、农家土鸡煲、金牌红烧肉

旅游商品：各种精美工艺品

沿途公交：游5、南汤线、123路的阳山碑材站

咨询电话：025-84110582/84111108

票价：48元

营业时间：8:00—17:00

2、先祖遗迹　古猿人洞景区

南京古猿人洞是全国重点文物保护单位。专家考证，南京古猿人头盖骨距今已有58万~63万年，同时发现的还有十五种动物骨化石共计2 000余件，其种类之广和数量之多，是国内外所罕见，其中葛氏斑鹿和肿骨鹿骨骼化石更是在长江以南地区首次发现，此洞被考古专家赞为"古生物宝洞"。

葫芦洞猿人头骨与头盖骨化石的发现被评为年度和"八五"期间全国十大考古发现，和北京猿人处于同时期，恰好一南一北遥相呼应，证实长江流域是中华民族的发祥地之一，更印证了江南人类文明源于汤山。

特色旅游项目：拜古猿人、观钟乳石

旅游商品：各种精美工艺品

沿途公交：南汤线、123路的汤山溶洞站

咨询电话：025-84110582/84111108

票价：25元

营业时间：8:00—17:00

Main Scenic Spots

1. Unique Massif——Yangshan Massif (Ming Cultural Villages)

Yangshan Massif Scenic Spot(Ming Cultural Villages), a National AAAA-Level Scenic Area and one of provincial cultural relics protection units, boasts of three sightseeing districts, including Ming Cultural Villages featured by making massif scene, Yangshan Limestone Stratum and world's greatest Yangshan Massif.

Food specialty: Yangshan Tofu, Stewed Chicken, Brand Pork Braised in Brown Sauce
Tourist commodities: varieties of exquisite handcrafts
Traffic : 5 Line, Nantang Line, 123 Line at Yangshan Massif Station
Telephone number: 025-84110582/84111108
Ticket price: ¥48
Business hours: 8:00—17:00

2.Relics of Progenitors——Troglodyte Cave Scenic Spot

Nanjing Troglodyte Cave is listed as one of the national key cultural relics protection units. Experts discovered that skulls of Nanjing cavemen had a history of 580,000 to 630,000 years. Meanwhile over 15 kinds of animal bone fossils were discovered as well, totaling more than 2,000 pieces, which was rare for its various kinds and large quantities at home and abroad. Bone fossils of Pseudaxis grayi and Megaloceros, in particular, were first discovered in the south of the Yangtze River. Therefore, the cave was named "Treasure Cave with Ancient Animals" by the archaeologists.

The discovery of skulls of ape-men in Hulu Cave was appraised one of National Top Ten Archaeological Discoveries that year and during the "Eight five" period as well. The cavemen in Hulu and Peking Man existed in the same period and corresponded to each other in the south and the north respectively, which proved that Yangtze River is one of the birthplaces of Chinese civilization and verified that the civilization of south of Yangtze River originated from Tangshan.

Special tourist activities: visiting ape-man, appreciating stalactites
Tourist commodities: varieties of exquisite handcrafts
Traffic: Nantang Line, 123 Line at Tangshan Cave Station
Telephone number: 025-84110582/84111108
Ticket price: ¥25
Business hours: 8:00—17:00

周边旅游

1、佛教圣迹——摩崖石刻

建于永乐初年，有"小千佛岩"和"江南第二云岗"之称。

2、旷世宝刹延祥寺、藏龙寺、隆昌寺（宝华山）

早在1996年就被林业部批准为国家森林公园，乾隆皇帝六次下江南"六登宝华山"。

3、风光旖旎姊妹湖——安基湖、汤泉湖

安基湖有"小九寨沟"之称。

4、民国印记——蒋介石温泉别墅

原系国民党元老张静江的私人别墅，1927年，他将此别墅送给了新婚燕尔的蒋介石、宋美龄夫妇，后成为蒋氏夫妇的私人专用别墅。

票价：12元
营业时间：8:00—17:00
咨询电话：025-68582668

Surrounding Tourism Areas

1. Buddhism Relics——Moya Inscription

It is built in the first of Yongle reign, also called "Small Thousand Buddha Rock" or "the Second Yungang of the South Region of Yangtze River".

2. Outstanding Temples——Yanxiang Temple, Canglong Temple, Longchang Temple

It was approved as National Forest Park by the Ministry of Forestry in 1996. The Emperor Qianlong climbed Baohua Mountain six times in the history.

3. Picturesque Sister Lakes——Anji Lake, Tangquan Lake

Anji Lake is also called "Small Jiuzhai Gou".

4. Impression of Republic of China —— Chiang Kai-shek's Villa with Spa

It was originally the private villa of Zhang Jingjiang, the senior statesman of the Kuomintang. In 1927, he presented it to the Mr. and Mrs. Chiang Kai-shek who got married and later it became their own villa with exclusive use.

Ticket price: ¥12
Business hours: 8:00—17:00
Telephone number: 025-68582668

酒店

1、香樟华苹温泉度假酒店

香樟华苹有东方风格、爪哇风格、马来风格、巴厘岛风格等六种不同风格的21幢独立院落式客房，每幢都配有私人泳池、露天的温泉按摩池，更有来自印尼的SPA服务。

咨询电话：025-84107777

Hotels

1. Kayumanis Nanjing Private Villa & Spa

Kayumanis Nanjing Private Villa & Spa boasts of 21 villas with independent courtyards in 6 different architectural styles, including oriental style, Java style, Malay style and Bali style. Every villa features its own swimming pool, an open-air hot spring with massage, and the service of SPA from Indonesia.

Telephone number: 025-84107777

2、御庭汤山温泉度假酒店

御庭汤山温泉度假酒店是世界小型豪华精品酒店会员单位，同时也是2007年度中国十佳旅游度假酒店之一。酒店拥有100间客房，提供传统泰式水疗和健康按摩美容。

咨询电话：025-87131188

3、御豪汤山温泉国际酒店

南京御豪汤山温泉国际酒店是民国建筑风格的温泉度假酒店。酒店拥有典雅舒适的各类客房164间套，专设行政楼层与温泉套房外，3套不同装修风格的豪华别墅更是招待贵宾政要的首选。酒店中西餐厅、宴会包厢、特色餐厅、咖啡厅、高档酒吧等餐饮场地。酒店温泉洗浴中心、健身房、游泳池、网球场、迷你高尔夫、桌球室、KTV、棋牌室等都是您放松身心、健康休闲的理想选择。

咨询电话：025-84109999

4、颐尚温泉度假村

颐尚温泉度假村是融度假酒店、温泉疗养、休闲娱乐为一体的五星级温泉度假村，有花瓣浴、养生浴、土耳其浴、芬兰浴、罗马大理石浴等共50个露天温泉池和15间高档雅致的私密性特色汤屋。

咨询电话：025-84103008、51190666

5、紫清湖生态旅游温泉度假村

紫清湖生态旅游温泉度假村融餐饮、客房、多功能会议室、观景温泉浴、山地迷你高尔夫、室内水上高尔夫练习、草地网球场等休闲娱乐商务于一体。客房数量为13间。

咨询电话：025-87127777、87138999

6、巴厘·原墅温泉会馆

原墅温泉会馆坐落于半山之上，群山环抱，绿树掩映，视野开阔，美景迷人，温泉喷涌，是风靡世界半山原生态的代表作。温泉SPA水疗区，内有温泉泡汤池、温泉游泳池、桑拿蒸房、健身房、按摩房。温泉泡汤池分别为冷泉池、气泡池、冲击池、浮浴池以及温泉按摩池。

咨询电话：025-84101771、84101772

2. Regalia Resort & SPA (Tangshan, Nanjing)

Regalia Resort & SPA is the member of world's small luxury hotels and one of Top Ten Holiday Resorts in 2007 as well. The hotel offers 100 rooms and provides the service of traditional Thai style spa and massage.

Telephone number: 025-87131188

3. Tangshan Yuhao Hot-spring International hotel

Tangshan Yuhao Hot-spring International Hotel features the architectural style of the Republic of China. The hotel has 164 various kinds of elegant and comfortable guest rooms with a floor of administration offices and suites with spa. Three sets of different decoration styles of luxury villas are the first choice for entertaining the VIPs. You can enjoy delicious food in a restaurant which offers Chinese and Western food, as well as in the specialty restaurant, cafe and high-end bar. You can recreate and relax yourself in a bathing center, a gym, a swimming pool, a tennis court, a mini golf course, a snooker room, a KTV room and a room for playing cards in the hotel.

Telephone number: 025-84109999

4. EA-Spring Resort & SPA

EA-Spring is a Five Star hotel integrating resorts, spa with recreation. It has 50 open-air hot-spring pools and 15 luxurious rooms with indoor hot-springs.

Telephone number: 025-84103008 51190666

5. Ziqing Lake Ecotourism Resort & SPA

Ziqing Lake Ecotourism Resort & SPA integrates catering, guestrooms, multifunction meeting rooms, spa with views, mountain-like mini golf course, indoor golf course in the water and tennis lawn, offering 13 guestrooms.

Telephone number: 025-87127777 87138999

6. Nanjing Vogue Hot-spring Hotel

Nanjing Vogue Hot-spring Hotel, located in the slope and surrounded by mountains, is one of the successes of world's popular originally-ecological hotels, with green trees, wide vision, beautiful scenery and gushing hot-springs. The area with spa is equipped with hot-spring bathing pools, hot-spring swimming pools, Sauna rooms, gyms and massage rooms. The hot-spring bathing pools include the cold-spring pools, bubble pools, and massage pools.

Telephone number: 025-84101771 84101772

娱乐

1、南京欢乐水魔方水上乐园

南京欢乐水魔方水上乐园以丰富多彩、惊险刺激的水上游乐设施为主的大型水上主题游乐园。园区分为激情冲浪区、魔法滑道区、儿童戏水区、歌舞表演区、休闲区、SPA水疗区六大区域。

咨询电话：025-84103566

乐园地址：南京市江宁区汤山街道黄栗墅

2、汤山翠谷现代农业科技园

南京汤山翠谷生态旅游度假村是观光休闲配套全的现代农业高科技生态观光园基础上建立而成的。园内视野开阔，布局合理，果茶林木成行成列，大棚设施规模宏大。宽广的葡萄架下果实累累，此起彼伏的山丘上葱绿满眼、果茶满坡。是以赏花、尝果、品茶、垂钓为主题的农业观光休闲园。

咨询电话：025-87161777

地址：南京市江宁区汤山街道上峰路1号

Recreation

1. Nanjing Happy Magic Watercube

Nanjing Happy Magic Watercube is a large water theme park, equipped with rich, colorful and adventurous aquatic amusement facilities. The park is divided into six areas, including passionate surfing area, magic slide area, children area, musical performance area, recreation area and spa area.

Telephone number:025-84103566

Address: Tangshan Street, Jiangning District, Nanjing

2. Tangshan Emerald Valley Agricultural Science and Technology Park

Tangshan Emerald Valley Ecotourism Resort & SPA is a modern agricultural high-tech ecological park, with beautiful scenery and various kinds of recreational facilities. The park is properly arranged with wide view and appropriate layout. It is an agricultural tourism and leisure park featuring seeing flowers, enjoying fruits, drinking tea and fishing.

Telephone number:025-87161777

Address: No.1 Shangfeng Road, Jiangning District, Nanjing

旅游路线

1、自驾游旅游线路

①上海方向到汤山

上海—沪宁高速—汤山出口—宁杭公路—南京古猿人洞—阳山碑材（明文化村）园区

②南京市区到汤山

a.中山门—沪宁高速—汤山出口—宁杭公路—南京古猿人洞—阳山碑材（明文化村）园区

b.中山门—宁杭公路—马群—麒麟—阳山碑材（明文化村）园区—南京古猿人洞

③杭州方向到汤山

杭州—宁杭高速—开城路出口—宁杭公路—阳山碑材（明文化村）园区—南京古猿人洞

2、公交线路

南汤线：南京火车站—阳山碑材—南京猿人洞

游5：南京火车站—阳山碑材

123路：麒麟悦民路—坟头站（阳山碑材园区）—南京猿人洞

金汤线：东山金宝市场—南京猿人洞

Tour Routes

1. Self-driving Travel

① From Shanghai to Tangshan

Shanghai—Huning Expressway—The Exit in Tangshan—Ninghang Highway—Nanjing Troglodyte Cave—Yangshan Massif (Ming Cultural Villages)

② From Nanjing Downtown to Tangshan

a. Zhongshan Gate—Huning Expressway—The Exit in Tangshan—Ninghang Highway—Nanjing Troglodyte Cave—Yangshan Massif (Ming Cultural Villages)

b. Zhongshan Gate—Ninghang Highway—Maqun—Qilin—Yangshan Massif (Ming Cultural Villages)—Nanjing Troglodyte Cave

③ From Hangzhou to Tangshan

Hangzhou—Ninghang Expressway—The Exit in Kaichenglu—Ninghang Highway—Yangshan Massif (Ming Cultural Villages)—Nanjing Troglodyte Cave

2. Buses

Nantang Line: Nanjing Railway Station—Yangshan Massif (Ming Cultural Villages)—Nanjing Troglodyte Cave

Travel 5 Line: Nanjing Railway Station—Yangshan Massif (Ming Cultural Villages)

No. 123 Line: Yuemin Road in Qilin Gate—Yangshan Massif (Ming Cultural Villages)—Nanjing Troglodyte Cave

Jintang Line: Jinbao Market in Dongshan County—Nanjing Troglodyte Cave

汤山方山国家地质公园
汤山园区导游图
Tourist Map of Tangshan District, Tangshan Fangshan National Geological Park

附录 Appendix

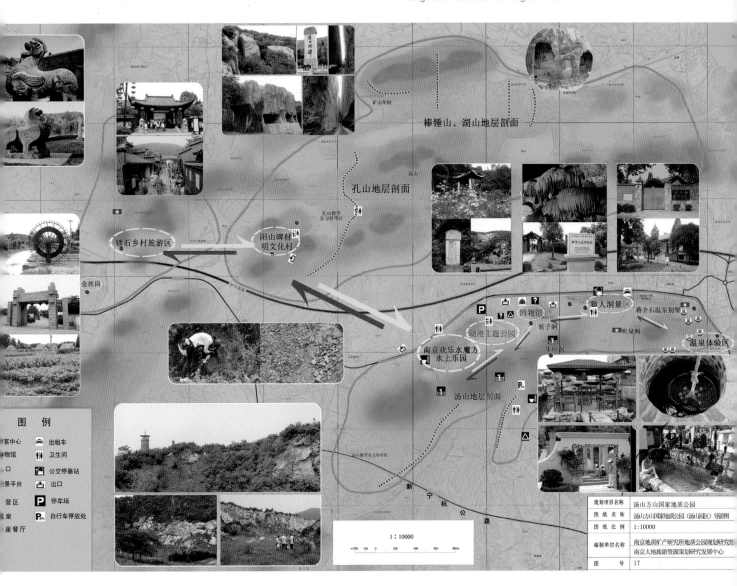

汤山园区地层简表
Strata Profile of Tangshan District

界	表	统	代号	地层单位	岩层厚（米）	地质年龄（亿年）	岩性简述
新生界	第四系	全新统	Q_4		0-67	0.01	灰白、褐黄色砾石、砂、黏土
		上更新统	Q_3	下蜀组	25	0.02	棕黄色粘土
		中更新统	Q_2	雨花台组	5.0	0.23	泥砂砾层、产雨花石
		下更新统	Q_1	三垛组	0.48-104	0.65	棕色砂泥岩
中生界	白垩系	上统	K_2	浦口组、赤山组	678	1.00	紫红色砂岩、砂砾岩、粉砂岩
		下统	K_1	扬冲组、上党组、圌山组（葛村组）	2174	1.45	流纹岩、粗面岩、集块岩等火山岩
	侏罗系	上统	J_3	西横山组、龙王山组、云合山组、大王山组	2000	1.61	下部为紫红色砂岩，上部为凝灰岩、安山岩
		中下统	J_{1+2}	象山组	380	1.76	长石石英砂岩为主，下部为砂砾岩层，上部页岩夹薄煤层
	三叠系	上统	T_3	范家塘组	225	2.00	细砂粉砂岩夹煤层
		中统	T_2	周冲村组、黄马青组	1469	2.37	紫红色细砂、粉砂岩
		下统	T_1	下青龙组、上青龙组	432	2.51	泥灰岩、白云岩、石膏层
古生界	二叠系	上统	P_2	龙潭组、大隆组	214	2.80	碳质页岩，粉砂岩含薄煤层
		下统	P_1	栖霞组、孤峰组	223	2.90	臭灰岩、燧石灰岩、硅质岩，砂页岩含磷结核，有蜓、珊瑚、菊石等化石
	石炭系	上统	C_3	船山组	40	3.06	灰白色夹黑色灰岩，含葛万藻和蜓类化石
		中统	C_2	黄龙组	84	3.18	质纯灰岩，底部为粗晶灰岩，含蜓类珊瑚等化石
		下统	C_1	金陵组、高骊山组、和州组、老虎洞组	96	3.59	灰岩、砂岩、泥灰岩、燧石等，含珊瑚、蜓类、腕足类、海百合茎等化石
	泥盆系	上统	D_3	五通组	70	3.83	石英砂岩，顶部为页岩，粘土岩
		中下统	D_{1-2}	观山组	100	4.28	中粗粒石英砂岩，底部石英砾岩
	志留系	上统	S_2	茅山组	26	4.36	茭缘，灰紫，紫红色粉砂岩、细砂岩
		下统	S_1	高家边组、侯家塘组、坟头组	1539	4.44	杂色粘土质粉砂岩，粘土岩，含三叶虫、笔石等化石
	奥陶系	上统	O_3	大田坝组、宝塔组、汤头组、五峰组	20	4.61	瘤状灰岩、泥灰岩、粘土岩，含三叶虫、笔石、腕足类等化石
		中统	O_2	汤山组	36	4.72	生物碎屑灰岩、泥灰岩，含直角石等化石
		下统	O_1	仑山组、红花园组、大湾组、牯牛潭组	295	4.88	灰岩、白云质灰岩，含牙形刺、鹦鹉螺、腕足类、三叶虫草、腹足类、苔藓虫化石
	寒武系	上统	ε_3	观音台组	625		白云岩、燧石灰岩